Health and Safety at Work:
Law and Practice

AUSTRALIA AND NEW ZEALAND
The Law Book Company Ltd.
Sydney : Melbourne : Perth

CANADA AND U.S.A.
The Carswell Company Ltd.
Agincourt, Ontario

INDIA
N. M. Tripathi Private Ltd.
Bombay
and
Eastern Law House
Calcutta and Delhi
M.P.P. House
Bangalore

ISRAEL
Steimatzky's Agency Ltd.
Jerusalem : Tel Aviv : Haifa

MALAYSIA : SINGAPORE : BRUNEI
Malayan Law Journal (Pte.) Ltd.
Singapore and Kuala Lumpur

Health and Safety at Work: Law and Practice

By

MICHAEL J. GOODMAN

M.A., Ph.D.
Solicitor; a Chairman of Industrial Tribunals;
a Social Security Commissioner

LONDON
SWEET & MAXWELL
1988

Published in 1988
by Sweet & Maxwell Limited of
11 New Fetter Lane, London
Computerset by Promenade Graphics Limited, Cheltenham
Printed in Scotland

British Library Cataloguing in Publication Data

Goodman, Michael J. (Michael Jack), *1931–*
  Health and safety at work: law and practice.
  1. Great Britain. Industrial health and
  safety. Law
  I. Title
  344.104'465

ISBN 0–421–37670–8

All rights reserved.
No part of this publication may be
reproduced or transmitted, in any form or by
any means, electronic, mechanical, photocopying,
recording or otherwise, or stored in any retrieval
system of any nature, without the written permission
of the copyright holder and the publisher, application
for which shall be made to the publisher.

©
Sweet & Maxwell
1988

Preface

This is a new book though some parts of it, with suitable modifications, originate in Sweet and Maxwell's Encyclopedia of Health and Safety at Work, Law and Practice, of which I am Editor. Entirely new are the chapters on Sources; Claims by Employees in Industrial Tribunals; and Claims by Employees, etc., for Social Security benefits, as well as the section in Chapter 2 on claims by Employees for Salary, Wages, Statutory Sick Pay, etc. The Appendices reproduce salient sections from the relevant Acts of Parliament and list the principal regulations, etc. All can be found printed in full, and annotated, in the Encyclopedia.

The original inspiration for this book was the Introductory section in the Health and Safety Encyclopedia. Indeed much of that Introduction, supplemented by additional material, is to be found reproduced in this book. I ought, therefore, at this stage to express gratitude to Bob Simpson, Senior Lecturer in Law at the London School of Economics and Political Science, for his considerable influence on the text of that Introduction in the years up to 1975, when I took over the Encyclopedia's editorship from him. He encouraged and helped me then and has now been kind enough to speak well of the prospects for this book.

I hope that this book will be of assistance to legal practitioners, personnel managers, trade union representatives, and all others who are concerned with the vital topic of health and safety at work. Dare I say, also that the law student may find here assembled materials that otherwise need some considerable library work?! I have tried for all of them, to combine comprehensiveness with readability.

I express my gratitude to all who have helped me with this book, whether my wife who has uncomplainingly put up with the "sidestresses" or the loyal and hard-working members of Sweet and Maxwell's staff, who have always performed the miracle of reducing my chaos into legal order.

Michael Goodman
May 1988

Contents

Preface v
Table of Cases ix
Table of Statutes xxvii
Table of Statutory Instruments xxxiii
Table of Abbreviations xxxv

1. INTRODUCTIONS AND SOURCES 1
   (1) Introductory 1
   (2) Source Materials 2

2. CLAIMS BY EMPLOYEES AND OTHERS IN THE COURTS 5
   (1) Claims for Damages for Diseases and Personal Injuries (at Common Law and for Breach of Statutory Duty) 5
   (2) Duty of Care 9
   (3) Standard of Care 12
   (4) Proof of Negligence 20
   (5) System of Work 24
   (6) Place of Work (Including Access) 27
   (7) Plant and Appliances 29
   (8) Protective Equipment 30
   (9) Negligence of Fellow Employee 33
   (10) Defences 34
   (11) Liability for Acts of Others 37
   (12) Damages 43
   (13) Time Limits 52
   (14) Compulsory Insurance 55
   (15) Claims by Employees for Salary, Wages, Statutory Sick Pay, etc. 56

3. LIABILITY OF EMPLOYER FOR DAMAGES FOR BREACH OF STATUTORY DUTY 61
   (1) Introductory 61
   (2) Standard of Care Required 68
   (3) Proof 72

(4) Defences	74
(5) Damages	78

4. LIABILITY OF EMPLOYER AS OCCUPIER — 79

5. LIABILITY FOR FATAL ACCIDENTS — 86
   (1) Introductory — 86
   (2) Survival of Causes of Action — 86
   (3) Liability to Dependants — 88
   (4) Interrelation of Statutory Rights — 95

6. CLAIMS BY EMPLOYEES IN INDUSTRIAL TRIBUNALS — 96

7. CLAIMS BY EMPLOYEES, ETC., FOR SOCIAL SECURITY BENEFITS — 101
   (1) Disablement Benefit — 102
   (2) Sickness and Invalidity Benefits — 105
   (3) Payments by the Department of Employment under the Pneumoconiosis, etc., (Workers' Compensation) Act 1979 — 105
   (4) System of Adjudication — 107

8. ADMINISTRATION AND ENFORCEMENT OF SAFETY, HEALTH AND WELFARE LEGISLATION — 110
   (1) Introductory — 110
   (2) Health and Safety Organisation — 111
   (3) Enforcement Powers — 114

APPENDIX 1 — 116
APPENDIX 2 — 152

Index — 157

Table of Cases

Addie & Sons Ltd. *v.* Dumbreck [1929] A.C. 358; 98 L.J.P.C. 119; 140 L.T. 650; 45 T.L.R. 267; 34 Com.Cas. 214, H.L. 84
Adsett *v.* K. & L. Steelfounders and Engineers [1953] 1 W.L.R. 773; 97 S.J. 419; [1953] 2 All E.R. 320; 51 L.G.R. 418, C.A. 70
—— *v.* West [1983] Q.B. 826; [1983] 3 W.L.R. 437; (1983) 127 S.J. 325; [1983] 2 All E.R. 985; (1983) 133 New L.J. 578 92
Allen *v.* Sir Alfred McAlpine and Sons Ltd. [1968] 2 Q.B. 229; 2 W.L.R. 366; [1968] 1 All E.R. 543, C.A. 54
Alford *v.* N.C.B. [1952] W.N. 144; [1952] 1 All E.R. 754, H.L. 40
Angus *v.* Glasgow Corp. 1977 S.L.T. 206 (Scot.) First Division 38
Arnold *v.* Central Electricity Generating Board (1987) 131 S.J. 167, C.A.; reversing [1986] 3 W.L.R. 171; (1986) 130 S.J. 484; (1986) 83 L.S.Gaz. 2090 52
Ashdown *v.* Williams (Samuel) and Son Ltd. [1957] 1 Q.B. 409; [1956] 3 W.L.R. 1104; 100 S.J. 945; [1957] 1 All E.R. 35, C.A. 83
Aspinall *v.* Sterling Mansell Ltd. [1981] 3 All E.R. 866 44
Aston *v.* Firth Brown [1984] 5 C.L. 142 20
Attnee *v.* Baker, *The Times*, November 18, 1983, D.C. 49
Austin *v.* Hart [1983] 2 A.C. 640; [1983] 2 W.L.R. 866; (1983) 127 S.J. 325; [1983] 2 All E.R. 341, P.C. 89

Bacon *v.* Jack Tighe (Offshore) and Cape Scaffolding [1987] 8 C.L. 195 34
Baker *v.* Dalgleish SS Co. [1922] 1 K.B. 361; 66 S.J. 318; 91 L.J.K.B. 392; 28 T.L.R. 245, C.A. 92
—— *v.* Hopkins (T.E.) & Son [1959] 1 W.L.R. 966; 103 S.J. 812; *sub nom.* Ward *v.* Hopkins (T.E.) & Sons [1959] 3 All E.R. 225 16
—— *v.* Market Harborough Industrial Co-operative Society [1953] 1 W.L.R. 1472; 97 S.J. 861, C.A. 21
—— *v.* Willoughby [1970] A.C. 467; [1970] 2 W.L.R. 50; [1969] 3 All E.R. 1528, H.L. 48
Bailey *v.* Barking and Havering Area Health Authority, *The Times*, July 22, 1978 94
—— *v.* Rolls Royce (1971) Ltd. [1984] 1 C.R. 688, C.A. 14
Barcock *v.* Brighton Corp. [1949] 1 K.B. 339; 65 T.L.R. 90; 93 S.J. 119; [1949] 1 All E.R. 251; 47 L.G.R. 253 .... 16, 74
Barkway *v.* South Wales Transport [1950] A.C. 185; 66 T.L.R. (Pt. 1 597); 114 J.P. 172; 94 S.J. 128; [1950] 1 All E.R. 391, H.L. 22
Barnes *v.* Bromley London Borough Council, *The Times*, November 19, 1983, D.C. 50, 102
—— *v.* Nayer, *The Times*, December 19, 1986, C.A. 6
Barr (E.S.) *v.* Cruickshank and Co. [1959] S.L.T. (Sh.Ct.) 9; 74 Sh.Ct. Rep. 218 10
Bastable *v.* Eastern Electricity Board [1956] 1 Lloyd's Rep. 586, C.A. 15

Table of Cases

Baxter v. St. Helena Group Hospital Management Committee, *The Times*, February 15, 1972 29
Bell v. Blackwood Morton & Sons, 1960 S.C. 11; 1960 S.L.T. 145 27
—— v. Secretary of State for Defence [1986] 2 W.L.R. 248; [1985] 3 All E.R. 661; (1985) 135 New L.J. 847; (1986) L.S.Gaz. 206, C.A. 8, 65
Benham v. Gambling [1941] A.C. 157; [1941] 1 All E.R. 7; 84 S.J. 703; 110 L.J.K.B. 49; 164 L.T. 290; 55 T.L.R. 177, H.L. 88
Bennett v. Chemical Construction (G.B.) Ltd. [1971] 1 W.L.R. 1571; 115 S.J. 550; [1971] 3 All E.R. 822, C.A. 21
Berry v. Humm & Co. [1915] 1 K.B. 627; 84 L.J.K.B. 918; 31 T.L.R. 198 92
—— v. Stone Manganese Marine Ltd. [1972] 1 Lloyd's Rep. 182; (1971) 115 S.J. 996; *The Times*, December 7, 1971 ..... 32, 67
Biddle v. Truvox Engineering Co. [1952] 1 K.B. 101; [1951] 2 T.L.R. 968; 95 S.J. 729; [1951] 2 All E.R. 835 64
Billings (A.C.) & Sons v. Riden [1958] A.C. 240; [1957] 3 W.L.R. 496; 101 S.J. 645; [1957] 3 All E.R. 1, H.L. 23
Birkett v. Hayes [1982] 1 W.L.R. 816; (1982) 126 S.J. 399; [1982] 2 All E.R. 70, C.A. 51
—— v. James [1978] A.C. 297; [1977] 3 W.L.R. 38; (1977) 121 S.J. 444; [1977] 2 All E.R. 801, H.L. [130 New L.J. 220; 77 L.S.Gaz. 1278, 330] 54, 55
Bishop v. Cunard White Star Co. Ltd. [1950] P. 240, p. 248 93
—— v. Starnes (J.S.) & Sons Ltd. and MacAndrews & Co. Ltd. [1971] 1 Lloyd's Rep. 162 83
Biss v. Lambeth, Southwark and Lewisham Area Health Authority (Teaching) [1978] 1 W.L.R. 382; (1978) 122 S.J. 32; [1978] 2 All E.R. 125, C.A. 55
Black v. Carricks (Caterers) Ltd. [1980] 1 R.L.R. 448, C.A. 26
Blake v. Midland Ry. (1852) 18 Q.B. 93; 21 L.J.Q.B. 233; 18 L.T.(O.S.) 330; 16 Jur. 562; 118 E.R. 35 94
Blundell v. Rimmer [1971] 1 W.L.R. 123; (1970) 15 S.J. 15; [1971] 2 Q.B. 262, C.A.; [1971] 1 All E.R. 1072 50
Board v. Hedley (Thomas) & Co. [1951] W.N. 422; [1951] 2 T.L.R. 779; 95 S.J. 546; [1951] 2 All E.R. 431, C.A. 13
Boden v. Moore (1961) 105 S.J. 510, C.A. 76
Bolt v. Moss and Sons Ltd. (1966) 110 S.J. 385 77
Bolton v. Stone [1951] A.C. 850; [1951] 1 T.L.R. 977; [1951] 1 All E.R. 1078; 50 L.G.R. 32; *sub nom.* Stone v. Bolton, 95 S.J. 333, H.L. 22
Bonnington Castings v. Wardlaw [1956] A.C. 613; [1956] 1 All E.R. 615, H.L.; 1956 S.C., H.L. 26 ..... 72, 73
Bontex Knitting Works v. St. John's Garage [1944] 1 All E.R. 381n.; 60 T.L.R. 253; affirming [1943] 2 All E.R. 690, C.A. 42
Boothman v. British Northrop Ltd. (1972) 13 K.I.R. 12, C.A. 34
Bowers v. Strathclyde Regional Council 1981 S.L.T. 122 49
Bowker v. Rose (1978) 122 S.J. 147; *The Times*, February 3, 1978, C.A. 49, 102
Boyle v. Kodak Ltd. [1969] 1 W.L.R. 661; 113 S.J. 382; [1969] 2 All E.R. 439; 6 K.I.R. 427, H.L. 76
—— v. Rennies of Dunfermline Ltd., 1975 S.L.T. (Notes) 13 44
Bradburn v. Great Western Ry. (1874) L.R. 10 Ex. 1 50
Bradford v. Boston Deep Sea Fisheries [1959] 1 Lloyd's Rep. 394 40
—— v. Robinson Rentals Ltd. [1967] 1 W.L.R. 337; 111 S.J. 33; [1967] 1 All E.R. 267 ..... 23, 45
Braham v. Lyons (J.) & Co. [1962] 1 W.L.R. 1048; 106 S.J. 588; [1962] 3 All E.R. 281; 60 L.G.R. 453, C.A. 64

British Airways Board v. Henderson [1979] I.C.R. 257 115
British Columbia Electric Ry. Co. v. Gentile (1914) 28 W.L.R. 795; 18 D.
  L.R. 264; 6 W.W.R. 1342; [1914] A.C. 1034; 111 L.T. 682 91
British Railways Board v. Herrington [1972] A.C. 877, H.L. 84
British Transport Commission v. Gourley [1956] A.C. 185; [1956] 2 W.L.R.
  41; 100 S.J. 12; [1955] 3 All E.R. 796; 49 R. & I.T. 11; [1955] 2 Lloyd's
  Rep. 475; [1955] T.R. 303; 34 A.T.C. 305 .... 48, 88
Brown v. Mills (John) & Co. (Llanidloes) Ltd. (1970) 114 S.J. 149, C.A. 17, 18
—— v. National Coal Board [1962] A.C. 574; [1962] 2 W.L.R. 269; 106 S.J.
  74; [1962] 1 All E.R. 81, H.L.; affirming [1961] 1 Q.B. 303, C.A.
  affirming [1960] 1 W.L.R. 67 70
—— v. Rolls Royce [1960] 1 W.L.R. 210; 104 S.J. 207; [1960] 1 All E.R. 577;
  1960 S.L.T. 119; 1960 S.C., H.L. 22 31
Bryce v. Swan Hunter Group, *The Times*, February 19, 1987 17
Buck v. English Electric Co. [1977] I.C.R. 629; [1977] 1 W.L.R. 806; [1978] 1
  All E.R. 271 53
Bunker v. Brand (Charles) Ltd. 2 Q.B. 80; [1969] 2 W.L.R. 1392; 113 S.J.
  487; [1969] 2 All E.R. 59; 6 K.I.R. 427 .... 80, 83
Burgess v. Florence Nightingale Hospital for Gentlewomen [1955] 1 Q.B.
  349; [1955] 2 W.L.R. 533; 99 S.J. 170; [1955] 1 All E.R. 511 92
—— v. Thorn Consumer Electronics (Newhaven) Ltd., *The Times*, May 16,
  1983 16
Burns v. British Railways Board [1977] 10 C.L. (unreported) C.A. 35
—— v. Terry [1951] 1 K.B. 454; [1951] 1 T.L.R. 349; 114 J.P. 613; 94 S.J.
  837; [1950] 2 All E.R. 987; 49 L.G.R. 161, C.A. 66, 69
Bushy v. Robert Watson (Constructional Engineers) Ltd. (1973) 13 K.I.R.
  498 27
Bux v. Slough Metals Ltd. [1973] 1 W.L.R. 1358; [1974] 1 All E.R. 262; 15
  K.I.R. 126, C.A.; 14 K.I.R. 179; 117 S.J. 615; [1974] 1 Lloyd's Rep.
  155 .... 61, 71
Byers v. Head, Wrightson & Co. [1961] 1 W.L.R. 961; 105 S.J. 569; [1961] 2
  All E.R. 538; 59 L.G.R. 385 76

Cackett v. Earl, *The Times*, October 19, 1976; [1976] C.L.Y. 666 49
Callaghan v. Kidd (Fredd) & Son [1944] K.B. 560; 113 L.J.K.B. 381; 170 L.T.
  368; 60 T.L.R. 409; 88 S.J. 239; [1944] 1 All E.R. 525, C.A. 35
Campbell v. Wallsend Slipway and Engineering Co. Ltd. [1978] I.C.R. 1015,
  D.C. 114
Canadian Pacific Steamships v. Bryers [1958] A.C. 485; 3 W.L.R. 993; 101
  S.J. 957; [1957] 3 All E.R. 572, H.L.; *sub nom.* Bryers v. Canadian
  Pacific Steamships [1957] 2 Lloyd's Rep. 387 .... 63, 64
Canadian Pacific Ry. v. Lockhart [1942] A.C. 591; [1942] 2 All E.R. 464,
  P.C. 39
Carlin v. Helical Bar Ltd. (1970) 9 K.I.R. 154 46
Carragher v. Singer Manufacturing Co. Ltd., 1974 S.L.T. (Notes) 28 .... 32, 67
Carroll v. Barclay (Andrew) & Sons, [1948] A.C. 477; [1948] L.J.R. 1490; 64
  T.L.R. 384; 92 S.J. 555; (1948) 2 All E.R. 386; 1948 S.C., H.L. 100;
  1948 S.L.T. 464 67
Cassidy v. Minister of Health [1951] 2 K.B. 343; [1951] 1 T.L.R. 539; 95 S.J.
  253; [1951] 1 All E.R. 574 .... 10, 43
Castanho v. Brown & Root (U.K.) Ltd. and Another [1980] 3 W.L.R. 991;
  124 S.J. 884; [1981] 1 All E.R. 143, H.L.; affirming [1980] 1 W.L.R.
  883; 124 S.J. 375, C.A.; reversing [1980] 1 All E.R. 689 6

Table of Cases

Caswell v. Powell Daffryn Associated Collieries [1940] A.C. 152; 108 L.J.K.B. 779; 161 L.T. 374; 55 T.L.R. 1004; 83 S.J. 976; [1939] 3 All E.R. 722, H.L. 34, 35
Caulfied v. Pickup [1941] 2 All E.R. 510 18
Century Insurance Co. v. Northern Ireland R.T.B. [1942] A.C. 509; 111 L.J.P.C. 138; 167 L.T. 404; [1942] 1 All E.R. 491, H.L. 42
Chapman v. Oakleigh Animal Products Ltd. (1970) 114 S.J. 432, C.A. 33
Charlton v. Forest Printing Ink Co. Ltd. [1980] I.R.L.R. 331; (1978) 122 S.J. 730 11
Chipchase v. British Titan Products Co. [1956] 1 Q.B. 545; [1956] 2 W.L.R. 677; 100 S.J. 186; [1956] 1 All E.R. 613; 54 L.G. R. 212, C.A. 18, 28
Chrysler (U.K.) Ltd. v. McCarthy [1978] I.C.R. 939, D.C. 115
Clarke v. Rotax Aircraft Equipment Ltd. [1975] 3 All E.R. 794, C.A. 51
Clifford v. Challeen (Charles) & Son [1951] 1 K.B. 495; [1951] 1 T.L.R. 234; 1 All E.R. 72, 391, C.A. 16, 31, 33, 67
Close v. Steel Co. of Wales [1962] A.C. 367; [1962] A.C. 367; [1961] 3 W.L.R. 319; [1961] 2 All E.R. 953, H.L. 12
Clowes v. N.C.B., *The Times*, April 23, 1987, C.A. 20
Coddington v. International Harvester Co. of G.B. Ltd. (1969) 113 S.J. 265; 6 K.I.R. 146 33
Cole v. Crown Poultry Packers Ltd. [1983] 3 All E.R. 226, C.A. 92
Colfar v. Coggins & Griffith (Liverpool) Ltd. [1945] A.C. 197, H.L. 27
Coltman v. Bibby Tankers, The Derbyshire [1987] 3 All E.R. 1068, H.L. 11, 30
Coltness Iron Co. v. Sharp [1938] A.C. 90; 106 L.J.P.C. 142; 157 L.T. 394; 53 T.L.R. 978; [1937] 3 All E.R. 593; 1937 S.C., H.L. 68; 1937 S.L.T., H.L. 589 70
Colvilles Ltd. v. Devine [1969] 1 W.L.R. 475; 113 S.J. 287; [1969] 2 All E.R. 53; 1969 S.C., H.L. 69 22
Compton v. McClure [1975] I.C.R. 378 40
Conway v. Wimpey (G.) & Co. (No. 2) [1951] 2 K.B. 266; [1951] 1 T.L.R. 587; 95 S.J. 156; [1951] 1 All E.R. 363 39
Cook v. Broderip (1968) 112 S.J. 193; 206 E.G. 128; 118 New L.J. 228 81
—— v. Keir (J.L.) & Co. Ltd. [1970] 1 W.L.R. 774; 114 S.J. 207; [1970] 2 All E.R. 518, C.A. 51
Cooke v. Midland G.W.Ry. of Ireland [1909] A.C. 229; 53 S.J. 319; 78 L.J.P.C. 76; 100 L.T. 626; 25 T.L.R. 375, H.L. 84
Cookson v. Knowles [1978] 2 W.L.R. 978; 122 S.J. 386; [1978] 2 All E.R. 604; affirming [1977] Q.B. 913; [1977] 3 W.L.R. 279; 121 S.J. 461; [1977] 2 All E.R. 820; [1977] 2 Lloyd's Rep. 412, C.A. 51, 93
Cooper v. Firth Brown Ltd. [1963] 1 W.L.R. 418; [1963] 2 All E.R. 31 49
Cork v. Kirby MacLean [1952] W.N. 399; [1952] 2 T.L.R. 217; 96 S.J. 482; [1952] 2 All E.R. 402; 50 L.G.R. 632; varying [1952] 1 All E.R. 1064 .... 9, 14
Coupland v. Arabian Gulf Oil Company (1983) 1 W.L.R. 1136; [1983] 3 All E.R. 226; 127 S.J. 393, C.A. 6
Cowhig v. Port of London Authority [1956] 2 Lloyd's Rep. 71
Crawley v. Gracechurch Line Shipping Ltd. [1971] 2 Lloyd's Rep. 179 82
—— v. Mercer, *The Times*, March 4, 1984 49
Croke v. Wiseman [1982] 1 W.L.R. 71; [1981] 3 All E.R. 853, C.A. 45, 46, 47
Crook v. Derbyshire Stone [1956] 1 W.L.R. 432; 100 S.J. 302; [1956] 2 All E.R. 447 40
Crookall v. Vickers-Armstrong [1955] 1 W.L.R. 659; 99 S.J. 401; [1955] 2 All E.R. 12; 53 L.G.R. 407 70
Curran v. Scottish Gas Board, 1970 S.L.T. (Notes) 33 23
Curwen v. James [1963] 1 W.L.R. 748; 107 S.J. 314; [1963] 2 All E.R. 619, C.A. 47

Table of Cases xiii

Cuther v. Vauxhall Motors Ltd. [1971] 1 Q.B. 418; [1970] 2 W.L.R. 961; 114
S.J. 247; [1970] 2 All E.R. 56 48

Daish v. Wauton [1972] 2 Q.B. 262; [1971] 2 W.L.R. 29; [1972] 1 All E.R. 25,
C.A.; *The Times*, October 16, 1971 50
Darby v. G.K.N. [1986] 1 C.R. 1 28
Davidson v. Handley Page (1945) 114 L.J.K.B. 81; 172 L.T. 38; 61 T.L.R.
178; 89 S.J. 118; [1945] 1 All E.R. 235, C.A. 28
Davie v. New Merton Board Mills [1959] A.C. 604; [1959] 21 W.L.R.
331; [1959] 1 All E.R. 346, H.L. 9, 11, 12, 30
Davies v. De Havilland Aircraft Co. [1951] 1 K.B. 50; 124 S.J. 535; [1950] 2
All E.R. 582; 49 L.G.R. 21 28
—— v. Powell Duffryn Associated Collieries Ltd. (No. 2) [1942] A.C. 601;
86 S.J. 294; [1942] 1 All E.R. 657; 111 L.J.K.B. 418; 167 L.T. 74; 58
T.L.R. 240 95
—— v. Taylor [1972] 3 W.L.R. 801; [1972] 3 All E.R. 836, H.L.(E.) 90
Davis v. Ministry of Defence, *The Times*, August 7, 1985, C.A. 53
Davison v. Apex Scaffolds [1956] 1 Q.B. 551; [1956] 2 W.L.R. 636; 100 S.J.
169; [1956] 1 All E.R. 473; 54 L.G.R. 171, C.A. 76
Denham v. Midland Employers Mutual Assurance [1955] 2 Q.B. 437; [1955] 3
W.L.R. 84; 99 S.J. 417; [1955] 2 All E.R. 561; [1955] 1 Lloyd's Rep.
467 9
Denman v. Essex Area Health Authority; Haste v. Sandell Perkins Ltd.
[1984] 3 W.L.R. 73; *The Times*, January 12, 1984 .... 50, 102
Dews v. N.C.B. [1987] 3 W.L.R. 38; *The Times*, June 8, 1987, H.L. 45
Dexter v. Courtaulds Ltd. [1984] 1 W.L.R. 372; 128 S.J. 81; [1984] 1 All E.R.
70, C.A. 51
Deyong v. Shenburn [1946] K.B. 227; 115 L.J.K.B. 262; 174 L.T. 129; 62
T.L.R. 193; 99 S.J. 139; [1946] 1 All E.R. 226 10
Dixon v. Cementation Ltd. [1960] 3 All E.R. 417, C.A. 27
Dodds v. Dodds [1978] Q.B. 543; [1978] 2 W.L.R. 434; (1977) 121 S.J. 619; 2
All E.R. 539 90
Dolbey v. Godwin [1955] 1 W.L.R. 553; 99 S.J. 335; [1955] 2 All E.R. 166,
C.A. 93
Donaghey v. Boulton & Paul Ltd. [1968] 1 A.C. 1; [1967] 3 W.L.R. 829; 111
S.J. 517; [1967] 2 All E.R. 1014; 2 K.I.R. 787 42
—— v. O'Brien (P.) & Co. [1966] 1 W.L.R. 1170; 110 S.J. 444; [1966] 2 All
E.R. 822; 1 K.I.R. 214, C.A. 76
Donovan v. Cammell Laird [1949] 2 All E.R. 82; 82 Lloyd's Rep. 642 25
Dooley v. Cammell Laird [1951] 1 Lloyd's Rep. 271 75
Doughty v. Turner Manufacturing Co. Ltd. [1964] 1 Q.B. 518; [1964] 2
W.L.R. 518; 108 S.J. 53; [1964] 1 All E.R. 98 .... 23, 45
Douglas C.B. v. Douglas and Another, 1974 S.L.T. (Notes) 67 44
Drennan v. Brooke Marine Ltd., *The Times*, June 21, 1983 .... 20, 32
Drummond v. British Building Cleaners [1954] 1 W.L.R. 1434; 98 S.J. 819;
[1954] 3 All E.R. 507; 53 L.G.R. 29, C.A. 19, 25, 73
Duffy v. Thanet District Council (1984) 134 New L.J. 680 40
Dunbar v. Painters (A. & B.) Ltd., *The Times*, March 14, 1986, C.A. 55
D'Urso v. Sanson [1939] 4 All E.R. 26; 83 S.J. 850 .... 28, 36
Dutton and Clark v. Daly [1985] I.C.R. 780; [1985] I.R.L.R. 363, E.A.T. 11, 98

Earle v. Medhurst [1985] 12 C.L. 296 43
East v. Beavis Transport Ltd. [1969] 1 Lloyd's Rep. 302, C.A. 39
Eaves v. Morris Motors [1961] 2 Q.B. 385; [1985] 3 W.L.R. 657; 105 S.J.
610; [1961] 3 All E.R. 233; 59 L.G.R. 466, C.A. 68

xiv   Table of Cases

Ebbs v. Whitson (James) [1952] 2 Q.B. 877; [1952] 1 T.L.R. 1428; 96 S.J. 375; [1952] 2 All E.R. 192, C.A. 67
Edwards v. N.C.B. [1949] 1 K.B. 704; 65 T.L.R. 430; 93 S.J. 337; [1949] 1 All E.R. 743, C.A. 70, 75
—— v. Stroud Riley & Co. [1984] 7 C.L. 309a 32
—— v. West Herts Group Hospital Management Committee [1957] 1 W.L.R. 415; 121 J.P. 212; 101 S.J. 190; [1957] 1 All E.R. 541 10
Excelsior Wire Rope Co. v. Callan [1930] A.C. 404; 99 L.J.K.B. 380; 142 L.T. 531; 94 J.P. 174; 35 Com.Cas. 300; 28 L.G.R. 543, H.L. 84

Farmer v. N.C.B., *The Times*,, April 27, 1985, C.A. 53
Farrell v. Rotary Services [1985] 1 N.I.J.B. 55 43
Ferguson v. Dawson (John) and Partners (Contractors) [1976] 1 W.L.R. 1213; 120 S.J. 603; [1976] 3 All E.R. 817; [1976] 2 Lloyd's Rep. 669; [1976] I.R.L.R. 346, C.A. 7
—— v. Home Office, *The Times*, October 8, 1977 .... 7, 63
—— v. Welsh and Others [1987] 3 All E.R. 777, H.L. 37
Field v. Perry's (Ealing) [1950] W.N. 320; [1950] 2 All E.R. 521 28
Finch v. Telegraph Co. [1949] W.N. 57; 65 T.L.R. 153; 93 S.J. 219; [1949] 1 All E.R. 452; 47 L.G.R. 710 .... 32, 33
Firman v. Ellis [1978] Q.B. 886; [1978] 3 W.L.R. 1; 122 S.J. 147; [1978] 2 All E.R. 702 53
Fisher v. C.H.T. (No. 2) Ltd. [1966] 2 Q.B. 475; [1966] 2 W.L.R. 391; 109 S.J. 933; [1966] 1 All E.R. 88, C.A.; reversing in part 236 L.T. 349 80
Fitzgerald v. Lane, *The Times*, March 7, 1987, C.A.; [1987] 2 All E.R. 455, C.A. 8, 35
Flower v. Ebbw Vale Steel, Iron & Coal Co. [1934] 2 K.B. 132; 103 L.J.K.B. 465; 151 L.T. 87; 78 S.J. 154 35
Fookes v. Slaytor [1978] 1 W.L.R. 1293; 122 S.J. 489 34
Forte's Service Areas Ltd. v. Department of Transport (1985) 31 Build.L.R. 1; (1986) 2 Const.L.J. 105, C.A. 38
Foster v. Tyne and Wear County Council [1986] 1 All E.R. 567, C.A. 49
Fowler v. British Railways Board, 113 S.J. 148; [1969] 1 Lloyd's Rep. 231, C.A. 27
Franklin v. Gramophone Co. [1948] 1 K.B. 542; [1948] L.J.R. 870; 64 T.L.R. 186; 92 S.J. 124; [1948] 1 All ER. 353 18

Galashiels Gas Co. v. O'Donnell (or Millar) [1949] A.C. 275; [1949] L.J.R. 540; 65 T.L.R. 76; 93 S.J. 71; [1949] 1 All E.R. 319; 47 L.G.R. 213; 1949 S.L.T. 223; *sub nom.* O'Donnell (or Millar) v. Galashiels Gas. Co., 1949 S.C., H.L. 31; 1949 S.L.T. 223 .... 72, 74, 75
Gallagher v. Balfour Beattie Co. 1951 S.C. 712 74
Galt v. British Railways Board (1983) 133 New L.J. 870 46
Gammell v. Wilson [1982] A.C. 27; [1981] 2 W.L.R. 248; 125 S.J. 116; [1981] 1 All E.R. 578, H.L.; affirming [1980] 3 W.L.R. 591; 124 S.J. 329; [1980] 2 All E.R. 557, C.A. 88
Gardiner v. Motherwell Machinery & Scrap Co. [1961] 1 W.L.R. 1424; [1961] 3 All E.R. 831, H.L.; 1962 S.L.T. 2; 1961 S.C., H.L. 1 .... 22, 73
Garard v. Southey (A.E.) and Co. [1952] 2 Q.B. 174; [1952] 1 T.L.R. 630; 96 S.J. 166; [1952] 1 All E.R. 597 .... 9, 41
Gavin v. Wilmot Breeden Ltd. [1973] 1 W.L.R. 1107; [1973] 3 All E.R. 935, C.A. 93
Gawtry v. Waltons Wharfingers & Storage Ltd. [1971] 2 Lloyd's Rep. 489, C.A. 13
Gemmell v. D.L.A. [1955] 1 Lloyd's Rep. 5, C.A. 28

General Cleaning Contractors v. Christmas [1953] A.C. 180; [1953] 2 W.L.R.
6; 97 S.J. 7; [1952] 2 All E.R. 1110; 51 L.G.R. 109, H.L. 15, 25, 27, 36
George and Richard (The) (1871) L.R. 3 A. & E. 466 90
Gibb v. United Steel Companies [1957] 1 W.L.R. 668; 101 S.J. 393; [1957] 2
All E.R. 110 9
Gilfillan v. N.C.B. 1972 S.L.T. (Sh.Ct.) 39 27
Ginty v. Belmont Building Supplies [1959] 1 All E.R. 414 74
Graham v. Co-operative Wholesale Society [1957] 1 W.L.R. 511; 101 S.J.
267; [1957] 1 All E.R. 654; 55 L.G.R. 137 .... 17, 67
—— v. Dodds [1983] 1 W.L.R. 808; [1983] 2 All E.R. 953, H.L. 93
—— v. Scotts Shipsbuilding & Engineering Co., Ltd. 1963 S.L.T. (Notes)
78 35
Grant v. N.C.B. [1956] A.C. 649; [1956] 2 W.L.R. 752; 100 S.J. 224; [1956] 1
All E.R. 682; 1956 S.C., H.L. 48; 1956 S.L.T. 155 67
Graves v. Hall (J. & E.) [1958] 2 Lloyd's Rep. 100 29
Gray v. Vickers-Armstrong (Shipbuilders) Ltd. [1963] 1 Lloyd's Rep. 143 26
Green v. Fibreglass [1958] 2 Q.B. 245; [1958] 3 W.L.R. 71; 102 S.J. 472;
[1958] 2 All E.R. 521 .... 43, 81
Green (R.H.) and S.W. Ltd. v. British Railways Board, *The Times*, October
8, 1980 8
Greenhalgh v. British Railways Board [1969] 2 Q.B. 286; [1969] 2 W.L.R.
892; 113 S.J. 108; [1969] 2 All E.R. 114; [1969] J.P.L. 343; 209 E.G.
825, C.A. 81
Grioli v. Allied Building Co. Ltd., *The Times*, April 10, 1985, C.A. 18
Guasto v. Robinsons of Winchester Ltd., *The Times*, November 23, 1977 47

Hall v. Avon Area Health Authority (Teaching) [1980] 1 W.L.R. 481; 124
S.J. 293; [1980] 1 All E.R. 516, C.A. 44
Hamilton v. N.C.B. [1960] A.C. 633; [1960] 2 W.L.R. 313; 124 J.P. 141; 104
S.J. 106; [1960] 1 All E.R. 76; 1960 S.C. (H.L.) 1; 1960 S.L.T.
24 .... 66, 72, 75
Harper v. Gray & Walker (1985) 1 W.L.R. 1196; (1985) 129 S.J. 776; [1985] 2
All E.R. 507; (1985) 82 L.S.Gaz. 3532; [1984] C.I.L.L. 106; (1983) 1
Const.L.J. 46 .... 8, 38
Harris v. Brights Asphalt Contractors [1953] 1 Q.B. 617; [1953] 1 W.L.R.
341; 97 S.J. 115; [1953] 1 All E.R. 395; 51 L.G.R. 296 28
Harrison v. N.C.B. [1951] A.C. 639; [1951] 1 T.L.R. 1079; 95 S.J. 413; [1951]
1 All E.R. 1102; 50 L.G.R. 1 66
Hartley v. Mayoh & Co. [1954] 1 Q.B. 383; [1954] 1 W.L.R. 355; 188 J.P. 178;
98 S.J. 107; [1954] 1 All E.R. 375, C.A. 18, 63
Harvey v. O'Dell (R.G.), Galway, Third Party [1958] 2 Q.B. 78; [1958] 2
W.L.R. 473; 102 S.J. 196; [1958] 1 All E.R. 657; [1958] 1 Lloyd's Rep.
273 .... 38, 39, 40
Hasley v. South Bedford Council, *The Times*, October 18, 1983 27
Haste v. Sandell Perkins Ltd. *See* Denman v. Essex Area Health Authority.
Hayes v. N.E. British Road Services [1977] 7 C.L. 173 24
Heard v. Brymbo Steel Co. [1947] K.B. 692; [1948] L.J.R. 372; 177 L.T. 251;
80 Lloyd's Rep. 424, C.A. 64, 75
Hegarty v. Rye-Arc [1953] 1 Lloyd's Rep. 465 26
Herbert v. Shaw (Harold) [1959] 2 Q.B. 138; [1959] 2 All E.R. 189 63
Herton v. Blaw Knox Ltd. (1968) 112 S.J. 963; (1968) 6 K.I.R. 35 25
Heskell v. Continental Express [1950] W.N. 210; 94 S.J. 339; [1950] 1 All
E.R. 1033; 83 Lloyd's Rep. 438 22
Hewson v. Grimsby Fishmeal Co. [1986] 9 C.L. 213 34

Table of Cases

Hicks v. British Transport Commission [1958] 1 W.L.R. 493; 102 S.J. 307; [1958] 2 All E.R. 39, C.A. 36
Hill v. James Crowe (Cases) Ltd. [1978] I.C.R. 298; [1978] 1 All E.R. 812; [1977] 2 Lloyd's Rep. 450 65
Hillen & Pettigrew v. I.C.I. [1936] A.C. 65; 104 L.J.K.B. 473; 153 L.T. 403; 51 T.L.R. 532; 41 Com.Cas. 29; affirming [1934] 1 K.B. 455 38
Hindle v. Porritt (Joseph) & Sons Ltd. [1976] 1 All E.R. 1142 13
Hinz v. Berry [1970] 2 Q.B. 40, C.A. 46
Hogan v. Bentinck West Hartley Collieries [1949] W.N. 109; [1949] L.J.R. 865; [1949] 1 All E.R. 588, H.L. 23
—— v. P & O Steam Navigation Co. [1959] 2 Lloyd's Rep. 305 81
Holland v. British Steel Corp. 1974 S.L.T.(N.) 72 45
Holson v. East Berkshire Area Health Authority [1987] 3 W.L.R. 232, H.L. 22
Honeywill & Stein Ltd. v. Larkin Bros. Ltd. [1934] 1 K.B. 191 42
Horry v. Tate & Lyle Refineries Ltd. [1982] 2 Lloyd's Rep. 416; *The Times*, March 17, 1982 ..... 52, 55
Hosking v. De Havilland Aircraft Co. [1949] 1 All E.R. 540; 83 Lloyd's Rep. 11 ..... 42, 64, 75
Housecraft v. Burnett [1986] 1 All E.R. 332, *The Times*, June 7, 1985, C.A. 46
Howitt v. Heads [1973] 1 Q.B. 64; [1972] 2 W.L.R. 183; [1972] 1 All E.R. 491 93
Hudson v. Ridge Manufacturing Co. [1947] 2 Q.B. 348; [1957] 2 All E.R. 229 ..... 33, 41
Hughes v. Lord Advocate [1963] A.C. 837; 1963 S.C., H.L. 31; 1963 S.L.T. 150; [1963] 2 W.L.R. 779; 107 S.J. 232; [1963] 1 All E.R. 705 45
Hultquist v. Universal Pattern and Precision Engineering Co. Ltd. [1960] 2 Q.B. 467; [1960] 2 W.L.R. 886; 104 S.J. 427; [1960] 2 All E.R. 226, C.A. 49
Hurley v. Sanders [1955] 1 W.L.R. 470; 99 S.J. 291; [1955] 1 All E.R. 833 27
Hussain v. New Taplow Paper Mills Ltd. [1987] I.C.R. 28; [1987] 1 All E.R. 417, C.A. 50
Hutchinson v. London & North Eastern Ry.Co. [1942] 1 K.B. 481; 111 L.J.K.B. 369; 166 L.T. 228; 58 T.L.R. 174; [1942] 1 All E.R. 330 68
Hyland v. R.T.Z. (Barium Chemicals) [1975] I.C.R. 54 36

Ilkiw v. Samuels [1963] 1 W.L.R. 991; 107 S.J. 680; [1963] 2 All E.R. 879, C.A. 38
Imperial Chemical Industries v. Shartwell [1965] A.C. 656; [1964] 3 W.L.R. 329; 108 S.J. 578; [1964] 2 All E.R. 999, H.L. 37, 77
Iqbal v. London Transport Executive, *The Times*, June 7, 1973, C.A. 40

Jackman v. Corbett [1987] 2 All E.R. 699, C.A. 102
Jagdeo v. Smith Industries [1982] I.C.R. 47, E.A.T. 11
James v. Durkin (Civil Engineering Contractors), *The Times*, May 25, 1983 16
Janor v. Morris [1981] 1 W.L.R. 1389; [1981] 3 All E.R. 780, C.A. 54
Jayes v. IMI (Kynoch) Ltd. [1985] I.C.R. 155; (1984) 81 L.S.Gaz. 3180, C.A. 77
Jayne v. N.C.B. [1963] 2 All E.R. 220 70
Jefford v. Gee [1970] 2 Q.B. 130; [1970] 2 W.L.R. 702; 114 S.J. 206; [1970] 1 All E.R. 1202, C.A. 50, 51
Jenkins v. Richard Thomas & Baldwins Ltd. [1966] 1 W.L.R. 476; 110 S.J. 111; [1966] 2 All E.R. 15, C.A. 47
Jenner v. Allen West & Co. [1959] 1 W.L.R. 554; 103 S.J. 371; [1959] 2 All E.R. 115, C.A. 15, 29
Jerred v. Dent [1948] 2 All E.R. 104; 81 Ll.L.R. 412; 92 S.J. 415 75

Jobling v. Associated Dairies Ltd. [1981] 3 W.L.R. 155; 125 S.J. 481; [1981] 2 All E.R. 752, H.L.; affirming [1981] Q.B. 389; [1980] 3 W.L.R. 704; 124 S.J. 631; [1980] 3 All E.R. 769, C.A. 48
Johnson v. Cammell-Laird & Co. Ltd. [1963] 1 Lloyd's Rep. 237 31
—— v. Croggan & Co. [1954] 1 W.L.R. 195; 98 S.J. 63; *sub nom.* Johnson v. Croggon & Co. (1954) 1 All E.R. 121; 52 L.G.R. 80 .... 34, 74
Jones v. Jones [1983] 1 W.L.R. 901; [1983] 1 All E.R. 1037; 127 S.J. 171 47
—— v. Livox [1952] 2 Q.B. 608; [1952] 1 T.L.R. 1377; 96 S.J. 344 34
—— v. Minton Construction Ltd. (1974) 15 K.I.R. 309 7
—— v. N.C.B. [1957] 2 Q.B. 55; [1957] 2 W.L.R. 760; 101 S.J. 319; [1957] 2 All E.R. 155, C.A. 71
—— v. Richards [1955] 1 W.L.R. 444; 99 S.J. 277; [1955] 1 All E.R. 463 13
—— v. Searle (G.D.) Co. Ltd. 122 S.J. 435; [1978] 3 All E.R. 851, C.A. 54
—— v. Smith (A.E.) Coggins [1955] 2 Lloyd's Rep. 17, C.A. 15
Joseph v. Ministry of Defence, *The Times*, March 4, 1980 17
Joyce v. Yeomans [1981] 1 W.L.R. 549; [1981] 2 All E.R. 21, C.A. 51

Kandalla v. British European Airways Corp. [1981] Q.B. 158; [1980] 2 W.L.R. 730; *sub nom.* Kandalla v. British Airways Board (formerly British Airways Corp.) (1979) 123 S.J. 769; [1980] 1 All E.R. 341 87
Karruppan Bhoomidas v. Port of Singapore Authority [1978] 1 W.L.R. 189; [1978] 1 Lloyd's Rep. 330; [1978] 1 All E.R. 956; (1977) 121 S.J. 816, P.C. 38
Kay v. I.T.W. [1968] 1 Q.B. 140; [1967] 3 W.L.R. 695; 111 S.J. 351; [1967] 3 All E.R. 22; 3 K.I.R. 18 C.A. 39
Kealey v. Heard [1983] 1 W.L.R. 573; [1983] 1 All E.R. 973; [1983] I.C.R. 484 21
Kerridge v. Port of London Authority [1952] 2 Lloyd's Rep. 142 71
Ketteman v. Hansel Properties Ltd., *The Times*, January 23, 1987, H.L. 54
Keys v. Shoe Fayre Ltd. [1978] I.R.L.R. 476 .... 11, 98
Kilgollan v. Cooke (W.) & Co. [1956] 1 W.L.R. 527; 100 S.J. 359; [1956] 2 All E.R. 294, C.A. 17, 67
Kinsella v. Harris Lebus Ltd. (1964) 108 S.J. 14, C.A. 19
Kirkup v. British Rail Engineering Ltd. C.A. [1983] 1 W.L.R. 1165; [1983] 3 All E.R. 147, C.A.; affirming [1983] 1 W.L.R. 190; [1983] 1 All E.R. 855 .... 20, 32
Kondis v. State Transport Authority (1984) 154 C.L.R. 672 37
Kossinski v. Chrysler United Kingdom Ltd. (1974) 118 S.J. 97; (1974) 15 K.I.R. 225, C.A. 9

Lamb v. Camden London Borough Council [1981] 2 W.L.R. 1038; 125 S.J. 356; [1981] 2 All E.R. 408, C.A. 14
Latimer v. A.E.C. [1953] A.C. 643; [1953] 3 W.L.R. 259; 117 J.P. 387; 97 S.J. 486; [1953] 2 All E.R. 449; 51 L.G.R. 457 .... 12, 13, 72
Lavender v. Diamints [1949] 1 K.B. 585; [1949] L.J.R. 970; 65 T.L.R. 163; 93 S.J. 147; [1949] 1 All E.R. 532; 47 L.G.R. 231 .... 64, 74
Lazarus v. Firestone Tyne & Rubber Co. [1963] C.L.Y. 2372 27
Lee v. Dickinson (John) & Co., 110 L.J. 317; [1960] 5 C.L. 370 27
—— v. South West Thames Regional Health Authority [1985] 1 W.L.R. 845, C.A. 43
Lees v. Secretary for Social Services [1985] A.C. 930; [1985] 2 W.L.R. 805; (1985) 129 S.J. 316; [1985] 2 All E.R. 203; (1985) 135 New L.J. 437; (1985) 82 L.S.Gaz. 1944, H.L. 108
Letang v. Cooper [1965] 1 Q.B. 232, C.A. 6

Lewis v. High Duty Alloy [1957] 1 W.L.R. 632; 101 S.J. 373; [1957] 1 All E.R.
  740; 55 L.G.R. 241 16
Light v. Bourne & Hollingsworth [1963] C.L.R. 2412 27
Limi v. Camden and Islington Area Health Authority [1980] A.C. 174; [1979]
  3 W.L.R. 44; 723 S.J. 457; [1979] 2 All E.R. 910, H.L.; affirming
  [1978] 3 W.L.R. 895; 122 S.J. 508, C.A.; affirming (1977) 122 S.J. 82;
  The Times, December 8, 1977 .... 45, 49, 51
Lincoln v. Hayman and Another [1982] 1 W.L.R. 488; [1982] R.T.R. 336,
  C.A. 49
Lister v. Romford Ice Storage Ltd. [1957] A.C. 555, H.L. 38
Littlejohn v. Claney, 1974 S.L.T. (Notes) 68 44
Logan v. Uttlesford District Council, *The Times*, July 31, 1984, C.A. 38
Lolley v. Keylock (1984) 81 L.S.Gaz. 1518, C.A. 93
Lovell v. Blundells [1944] K.B. 502; 113 L.J.K.B. 385; 170 L.T. 323; 60
  T.L.R. 326; [1944] 1 All E.R. 53 29

MacKay v. Culsa Shipbuilding Co. 1945 S.C. 414; 1946 S.L.T 104 66
MacShannon v. Rockwear Glass Ltd. [1978] 1 All E.R. 625, H.L. 6
Mace v. Green (R. & H.) [1959] 2 Q.B. 14; [1959] 2 W.L.R. 504; 103 S.J. 293;
  [1959] 1 All E.R. 655; [1959] 1 Lloyd's Rep. 146 10
Maguire v. Lagan (P.J.) (Contractors) [1976] N.I. 49, C.A. 42
Mahoney v. Hay's Wharf [1963] 2 Lloyd's Rep. 312 28
Malcolm v. Broadhurst [1970] I.C.R. 60 45
Marsh v. Moones 2 K.B. 208; [1949] L.J.R. 1313; 65 T.L.R. 318; 113 J.P. 346;
  98 S.J. 450; [1949] 2 All E.R. 21; 47 L.G.R. 418 39
Marshall v. Gotham [1954] A.C. 360; [1954] 2 W.L.R. 812; 98 S.J. 268; [1954]
  1 All E.R. 937, H.L. 69, 72
Marston v. British Railways Board [1976] I.C.R. 124 (D.C.); [1976] I.C.R.
  353, C.A. 30
Massey-Harris-Ferguson v. Piper [1956] 2 Q.B. 396; [1956] 3 W.L.R. 271; 100
  S.J. 472; [1956] 2 All E.R. 722; 54 L.G.R. 410 63
Matthews v. Kuwait Bechtel [1959] 2 Q.B. 57; [1959] 2 W.L.R. 702; 103 S.J.
  393; [1959] 2 All E.R. 345 6
Mainz v. Dodd, *The Times*, July 21, 1978; (1978) 122 S.J. 645 54
McCann v. Sheppard [1973] 1 W.L.R. 540; [1973] 2 All E.R. 881, C.A. 90
McCafferty v. Metropolitan Police District Receiver [1977] I.C.R. 799;
  [1977] 1 W.L.R. 1073; 121 S.J. 678; [1977] 2 All E.R. 766, C.A. 32, 53
McCarthy v. Coldair [1951] W.N. 590; [1951] 2 T.L.R. 1226; 95 S.J. 711; 50
  L.G.R. 65 .... 66, 68
—— v. Daily Mirror Newspapers [1949] 65 T.L.R. 501; 113 J.P. 229; 93 S.J.
  237; [1949] 1 All E.R. 801; 47 L.G.R. 588 10
—— v. O'Flynn [1979] I.R. 127, Eire Sup.Ct. 43
McCloskey v. Western Health and Social Services Board [1983] 4 N.I.J.B.,
  C.A. 10
McDermid v. Nash Dredging & Reclamation Co. Ltd., *The Times*, July 31,
  1984; [1986] Q.B. 965, C.A.; [1987] 2 All E.R. 878; *The Times*, July 3,
  1987 .... 9, 10, 37, 51
McDonald v. B.T.C. [1955] 1 W.L.R. 1323; 99 S.J. 912; [1955] 3 All E.R. 789;
  [1955] 2 Lloyd's Rep. 467 28
McIvor and Another v. Southern Health & Social Services Board, Northern
  Ireland [1978] 1 W.L.R. 757; 122 S.J. 368; [1978] 2 All E.R. 625 43
McKean v. Raynor Bros. (1942) 167 L.T. 369; 86 S.J. 376; [1942] 2 All E.R.
  650 .... 38, 39
McKew v. Holland and Hannan & Cubitts (Scotland) Ltd., 1970 S.C., H.L.
  20; 1970 S.L.T. 68; [1969] 3 All E.R. 1621, H.L. 23, 48

McMullen v. National Coal Board [1982] I.C.R. 148 37
McPhee v. General Motors Ltd. (1970) 8 K.I.R. 885 29
McWilliams v. Sir William Arrol & Co. Ltd. [1962] 1 W.L.R. 295; 106 S.J.
   218; 1962 S.C.(H.L.) 70; 1962 S.L.T. 121; *sub nom.* Cummings (or
   McWilliams) v. Sir Williams Arrol & Co. [1962] 1 All E.R. 623; 1961
   S.C. 134; 1961 S.L.T. 265; affirmed 1962 S.C., H.L. 70; 1962 S.L.T.
   821, H.L. 74
Mead v. Clark, Chapman & Co. Ltd. [1956] 1 W.L.R. 76; 100 S.J. 51; [1956] 1
   All E.R. 44, C.A. 94
Meah v. McCreamer [1985] 1 All E.R. 367; (1985) 135 New L.J. 80 47
Mears v. Safecar Security [1983] Q.B. 54; [1982] I.C.R. 626; [1982] 3 W.L.R.
   366; [1982] 2 All E.R. 865; [1982] I.R.L.R. 183; (1982) 79 L.S.Gaz.
   921, C.A.; affirming [1981] I.C.R. 409; [1981] 1 W.L.R. 1214; [1981]
   I.R.L.R. 99, E.A.T. 57
Megarity v. Ryan (D.J.) & Sons [1980] 1 W.L.R. 1237; (1980) 124 S.J. 498;
   [1980] 2 All E.R. 832, C.A. 44
Mehmet v. Derry [1977] 2 All E.R. 529 92
Mersey Docks & Harbour Board v. Coggins & Griffith [1947] A.C. 1; [1946] 2
   All E.R. 345, H.L. 41, 42
Miller v. Ministry of Pensions [1947] W.N. 241; [1948] L.J.R. 203; 177 L.T.
   536; 63 T.L.R. 474; 91 S.J. 484; [1947] 2 All E.R. 372 20
Minter v. Contractors (D. & H.) (Cambridge) *The Times*, June 30, 1983 35
Money v. Thorn Electrical Industries [1977] 11 C.L. (unreported), C.A. 74
Moodie v. Furness Shipbuilding [1951] 2 Lloyd's Rep. 600, C.A. 69
Moore v. Fox [1956] 1 Q.B. 596; [1956] 2 W.L.R. 342; [1956] 1 All E.R. 182;
   [1956] 1 Lloyd's Rep. 129 21
Morgan v. Wallis (T.) Ltd. [1974] 1 Lloyd's Rep. 165 46
Morris v. Ford Motor Co. Ltd. [1973] 1 Q.B. 792; [1973] 2 W.L.R. 843; 117
   S.J. 393; [1973] 2 All E.R. 1084 .... 8, 11, 38
—— v. Martin (C.W.) Ltd. [1966] 1 Q.B. 716; [1965] 3 W.L.R. 276; 109 S.J.
   451; [1965] 2 All E.R., C.A. 41
—— v. West Hartlepool Steam Navigation Co. [1956] A.C. 552; [1956] 1
   W.L.R. 177; 100 S.J. 129; [1956] 1 All E.R. 385; [1956] 1 Lloyd's Rep.
   76 .... 13, 17
Mulholland v. Mitchell [1971] AC. 666, H.L.; [1971] 2 W.L.R. 93; (1970) 115
   S.J. 15; [1971] 1 All E.R. 307 47
—— v. William Reid & Leys, 1958 S.C. 290; 1958 S.L.T. 285 39
Mullard v. Ben Line Steamers Ltd. [1970] 1 W.L.R. 1414; [1971] 2 All E.R.
   424; [1970] 2 Lloyd's Rep. 121, C.A. 77
Mulready v. Bell [1953] 2 Q.B. 117; [1953] 3 W.L.R. 100; 97 S.J. 419; [1953] 2
   All E.R. 215, C.A. 75
Murfin v. United Steel Companies [1957] 1 W.L.R. 104; 101 S.J. 61; [1957] 1
   All E.R. 23; 55 L.G.R. 43 68
Murphy v. Stone Wallwork (Charlton) Ltd. [1969] 1 W.L.R. 1023; 113 S.J.
   546; [1969] 2 All E.R. 949, H.L. 47
Murray v. Schwachman [1938] 1 K.B. 130; 106 L.J.K.B. 354; 156 L.T. 407; 53
   T.L.R. 458; 81 S.J. 294; [1937] 2 All E.R. 68; 30 B.W.C.C. 466 71
—— v. Walnut Cabinet Works, 105 L.J. 41; [1954] C.L.Y. 1320, C.A. 28
Mustart v. Post Office, *The Times*, February 11, 1982 46

Nabi v. British Leyland (U.K.) Ltd. [1980] 1 W.L.R. 529; [1980] 1 All E.R.
   667, C.A. 49
Nance v. British Columbia Electric Ry. [1951] A.C. 601; [1951] 2 T.L.R. 137;
   95 S.J. 543; [1951] 2 All E.R. 448, P.C. 34
Nancollas v. Insurance Officer [1985] 2 All E.R. 833, C.A. 104

Napievalski v. Curtis (Contractors) [1959] 1 W.L.R. 835; 103 S.J. 695; [1959]
    2 All ER. 426; 58 L.G.R. 70 63
National Coal Board v. England [1954] A.C. 403; [1954] 2 W.L.R. 400; 98
    S.J. 176; [1954] 1 All E.R. 456 18
Nelson v. Duraplex Industries Ltd., 1975 S.L.T. (Notes) 31 44
Nethergate Ltd. v. Taverna, *The Times*, [1984] I.R.L.R. 240; (1984) 134 New
    L.J. 544, C.A. 7
Newall v. Tunstall [1971] 1 W.L.R. 105; (1970) 115 S.J. 14; [1970] 3 All E.R.
    465 50
Newman v. Harland & Wolff [1953] 1 Lloyd's Rep. 114 13
Nicholls v. Reemer 107 L.J. 378; [1947] C.L.Y. 2351 29
Nicholson v. Atlas Steel Foundry and Engineering Co. [1957] 1 W.L.R. 613;
    [1957] 1 All E.R. 776, H.L.; 1957 S.C., H.L. 44 .... 67, 72, 73
Nicolson v. Shaw Savill [1957] 1 Lloyd's Rep. 162 16
Nimmo v. Alexander Cowan & Sons [1968] A.C. 107; [1967] 3 W.L.R. 1169;
    111 S.J. 668; [1967] 3 All E.R. 187; 1967 S.L.T. 277; 1967 S.C., H.L.
    79; 3 K.I.R. 277 71
Nolan v. Dental Manufacturing co. [1958] 1 W.L.R. 936; 102 S.J. 619; [1958]
    2 All E.R. 449 .... 31, 48, 73
Norris v. Syndic Manufacturing Co. [1952] 2 Q.B. 135; (1952) 1 All E.R. 935,
    C.A. 33, 66, 71, 75
Nottingham v. Aldridge, Prudential Assurance Co. (Third Party) [1971]
    2 Q.B. 739; [1971] 3 W.L.R. 1; 10 K.I.R. 252; [1971] 2 All E.R. 751 41
Nunan v. Southern Ry. [1924] 1 K.B. 223; 68 S.J. 139; 93 L.J.K.B. 140; 130
    L.T. 131; 40 T.L.R. 21, C.A. 91
Nutbrown v. Rosier, *The Times*, March 1, 1982 92

Oliver v. Ashman [1962] 2 Q.B. 210; [1961] 3 W.L.R. 669; 105 S.J. 608; [1961]
    3 All E.R. 323, C.A.; reversing [1961] 1 Q.B. 337; [1960] 3 W.L.R.
    924; 104 S.J. 1036; [1960] 3 All E.R. 667; [1960] C.L.Y. 867 46
O'Reilly v. I.C.I. [1955] 1 W.L.R. 1155; 99 S.J. 778; [1955] 3 All E.R. 382 9
—— v. National Rail and Tramway Appliances Ltd. [1966] 1 All E.R. 499 33
Oropesa (The) [1943] 1 All E.R. 221, C.A. 23
O'Sullivan v. Herdmans Ltd., *The Times*, July 10, 1987, H.L. 43
Overseas Tankship (U.K.) Ltd. v. Morts Dock & Engieering Co. (The
    Wagon Mound) [1961] A.C. 388; [1961] 2 W.L.R. 126; 105 S.J. 85;
    [1961] 1 All E.R. 404; [1961] 1 Lloyd's Rep. 1 23

Paine v. Colne Valley Electricity Supply Co. (1938) 160 L.T. 124; 55 T.L.R.
    181; 83 S.J. 115; [1938] 4 All E.R. 803 *per* Goddard L.J. at 807 43
Palfrey v. Greater London Council [1985] I.C.R. 437 49
Paling v. Marshall (A.) (Plumbers) (November 22, 1957) unreported; [1957]
    C.L.Y. 2420 31
Paris v. Stepney Borough Council [1951] A.C. 367; [1951] 1 T.L.R. 25; 115
    J.P. 22; 94 S.J. 837; [1951] 1 All E.R. 42; 49 L.G.R. 293; 84 Lloyd's
    Rep. 525 .... 9, 12, 30
Parker v. Vickers Ltd. [1979] 10 C.L. 161 27
Parry v. Cleaver [1970] A.C. 1 H.L.; reversing [1968] 1 Q.B. 195, C.A. 50
Parsons v. B.N.M. Laboratories [1964] 1 Q.B. 95; [1963] 2 W.L.R. 1273; 107
    S.J. 294; [1963] 2 All E.R. 658 49
Pasfield v. Hay's Wharf [1954] 1 Lloyd's Rep. 150 26
Pass of Ballater (The) [1942] P. 112 at 117 42
Paterson v. Costain and Press (Overseas) Ltd. (1979) 123 S.J. 142; [1978] 1
    Lloyd's Rep. 86 .... 39, 40
Payne v. Peter Bennie Ltd. (1973) 14 K.I.R. 395 15

Payne v. Weldless Steel Tube Co. [1956] 1 Q.B. 196; [1955] 3 W.L.R. 771; 99 S.J. 814; [1955] 3 All E.R. 612; 54 L.G.R. 19 .... 71, 72
Pearce v. Secretary of State for Defence [1987] 2 W.L.R. 782; *The Times*, December 31, 1986; *The Times*, August 5, 1987, C.A. 8, 65
Peat v. Muschamp (N.J.) & Co. Ltd. (1970) 7 K.I.R. 469 at 476–477, C.A. 26
Pentney v. Anglican Water Authority [1983] I.C.R. 464 31
Perez v. C.A.V. Ltd. [1959] 1 W.L.R. 724; 103 S.J. 492; [1959] 2 All E.R. 414 49
Photo Production v. Securicor Transport [1980] A.C. 827; [1980] 2 W.L.R. 283; 124 S.J. 147; [1980] 1 All E.R. 556; [1980] 1 Lloyd's Rep. 545, H.L.; reversing [1978] 1 W.L.R. 856; 122 S.J. 315; [1978] 3 All E.R. 146; [1978] 2 Lloyd's Rep. 172, C.A. 41
Pickett v. British Rail Engineering Ltd. [1980] A.C. 136; [1978] 3 W.L.R. 955; [1979] 1 All E.R. 774; [1979] 1 Lloyd's Rep. 519, H.L.; reversing (1977) 121 S.J. 814, C.A. 45, 46, 51, 93
Pigney v. Pointer's Transport [1957] 1 W.L.R. 1121; 101 S.J. 851; [1957] 2 All E.R. 807 .... 23, 95
Pipe v. Chambers Wharf [1952] 1 Lloyd's Rep. 194 26
Plummer v. Wilkins and Son Ltd., *The Times*, July 9, 1980 49
Port Swettenham Authority v. T.W.Wu & Co. [1978] 3 W.L.R. 530; 122 S.J. 523; [1978] 3 All E.R. 337 41
Porteous v. National Coal Board [1967] S.L.T. 117 14
Portsea Island Mutual Co-operative Society Ltd. v. Leyland [1978] I.C.R. 1195; 122 S.J. 486; [1978] I.R.L.R. 556 39
Pratt v. Richards [1951] 2 K.B. 208; [1951] 1 T.L.R. 515; [1951] 1 All E.R. 90n. 28
Prince v. Carrier Engineering Co. [1955] 1 Lloyd's Rep. 401 .... 25, 28
Prescott v. Bulldog Tools Ltd. [1981] 3 All E.R. 869 44
Presho v. Department of Health and Social Security Insurance Officer [1984] A.C. 310; [1984] 2 W.L.R. 29; [1984] 1 All E.R. 97; [1984] I.C.R. 463; [1984] I.R.L.R. 74; (1984) 128 S.J. 18; (1984) 134 New L.J. 38; (1984) 81 L.S.Gaz. 436, H.L.; reversing [1983] I.C.R. 595; (1983) 127 S.J. 425; [1983] I.R.L.R. 295, C.A. 108
Pritchard v. Cobden (J.H.) Ltd. and Another, *The Times*, August 27, 1986 and December 3, 1986, C.A. (over-ruling Jones v. Jones [1985] Q.B. 794, C.A.) 47
Pullin v. Prison Commissioners [1957] 1 W.L.R. 1186; 101 S.J. 903; *sub nom.* Pullin v. Prison Commissioners [1957] 3 All E.R. 470 7
Pym v. Great Northern Ry. (1862) 2 B. & S. 759; 31 L.J.Q.B. 249; 6 L.T. 537; 8 Jur.N.S. 819; 10 W.R. 737; 121 E.R. 1254; affirmed (1863) 4 B. & S. 396 92

Qualcast (Wolverhampton) v. Haynes [1959] A.C. 743; [1959] 2 W.L.R. 510; 103 S.J. 310; [1959] 2 All E.R. 38, H.L. 18, 31, 33
Quinn v. Burch Bros. (Builders) Ltd. [1966] 2 Q.B. 370; [1966] 2 W.L.R. 1017; 110 S.J. 214; [1966] 2 All E.R. 283; 1 K.I.R. 9; 82 L.Q.R. 296, C.A. 7
—— v. Cameron & Robertson [1958] A.C. 9; [1959] 2 W.L.R. 692; 101 S.J. 317; [1957] 1 All E.R. 760; 55 L.G.R. 177; 1957 S.C., H.L. 22; 1957 S.L.T. 143; reversing 1956 S.C. 224 17

R. v. Industrial Injuries Commissioners *ex p.* A.E.U. (No. 2) [1966] Q.B. 31, C.A. 104
Rands v. McNeil [1955] 1 Q.B. 253; [1954] 3 W.L.R. 905; 98 S.J. 851; [1954] 3 All E.R. 593, C.A. 15

Rankine v. Garton Sons & Co. Ltd. [1979] 2 All E.R. 1185; 123 S.J. 305, C.A. 20
Rawlinson v. Babcock & Wilcox Ltd. [1967] 1 W.L.R. 481; 111 S.J. 76; *sub nom.* Moore v. Babcock & Wilcox; Rawlinson v. Same [1966] 3 All E.R. 882 94
Reading v. Harland & Wolff [1954] 1 Lloyd's Rep. 131, C.A. 28
Rees v. Bernard Hastie & Co. [1953] 1 Q.B. 328; [1953] 2 W.L.R. 288; 97 S.J. 94; [1953] 1 All E.R. 375; 51 L.G.R. 189 66
Reincke v. Gray [1964] 1 W.L.R. 832; 108 S.J. 461; [1964] 2 All E.R. 687, C.A. 94
Rialas v. Mitchell, *The Times*, July 17, 1984, C.A. 46
Richards v. Highway Ironfounders (West Bromwich) [1957] 1 W.L.R. 781; 99 S.J. 580; [1955] 3 All E.R. 205; 53 L.G.R. 641, C.A. 67, 70
Richardson v. Stephenson Clarke Ltd. [1969] 1 W.L.R. 1695; 113 S.J. 873; [1969] 3 All E.R. 705 29
Riddick v. Weir Housing Corp. Ltd. [1971] S.L.T. 24; 1970 S.L.T. (Notes) 71 17
Roberts v. Dorman Long & Co. [1953] 1 W.L.R. 942; 97 S.J. 487; [1953] 2 All E.R. 428; 51 L.G.R. 476 .... 71, 73, 74
—— v. Wallis (T.) and l'Italica di Navigazione, S.P.A. [1958] 1 Lloyd's Rep. 29 29
Robinson v. British Rail Engineering [1981] 8 C.L. 63, *The Times*, June 27, 1980, C.A. 32
—— v. Post Office (1973) 117 S.J. 915; [1974] 2 All E.R. 737, C.A. 23, 45
Roe v. Ministry of Health; Woolley v. Same [1954] 2 Q.B. 66; [1954] 2 W.L.R. 915; 98 S.J. 319; [1954] 2 All E.R. 131 21
Roles v. Nathan (T/A Manchester Assembly Rooms) [1963] 1 W.L.R. 1117; 107 S.J. 680; [1963] 2 All E.R. 908, C.A. 83
Ronex Properties Ltd. v. John Laing Ltd. [1982] 3 W.L.R. 875; 126 S.J. 727; [1982] 3 All E.R. 961, C.A. 8, 38
Rose v. Ford [1937] A.C. 826; 81 S.J. 683; [1937] 3 All E.R. 359; 106 L.J.K.B. 376; 157 L.T. 174; 53 T.L.R. 873, H.L. 88
—— v. Plenty [1976] 1 All E.R. 97; (1975) 119 S.J. 592, C.A. 39
Ross v. Associated Portland Cement Manufacturers Ltd. [1964] 1 W.L.R. 768; 108 S.J. 460; [1964] 2 All E.R. 452; 62 L.G.R. 513, H.L. 74, 76
Rothwell v. Caverswall Stone Co. (1944) 113 L.J.K.B. 520; 171 L.T. 289; 61 T.L.R. 17; [1944] 2 All E.R. 350; 37 B.W.C.C. 72 23
Roughead v. Railway Executive [1949] W.N. 280; 65 T.L.R. 435; 93 S.J. 514 93
Rourke v. Barton, *The Times*, June 23, 1982 46
Rowden v. Gosling [1953] C.P.L. 218; [1953] C.L.Y. 2420 28
Rudy v. Tay Textiles (O.H.) 1978 S.L.T. (Notes) 62 49
Ryan v. Maubre Sugars Ltd. (1970) 114 S.J. 492, C.A. 36

Salmon v. Seafarer Restaurant [1983] 1 W.L.R. 1264; [1983] 3 All E.R. 729 63
Salsbury v. Woodland [1970] 1 Q.B. 324, C.A.; 113 S.J. 327; [1969] 3 All E.R. 863 42
Sandford v. Eugene Ltd. (1971) 115 S.J. 33 82
Sanger v. Kent and Callow [1978] 11 C.L. 75 49
Selvanayagam v. University of West Indies [1983] 1 W.L.R. 585; [1983] 1 All E.R. 824, P.C. 48
Semtex Ltd. v. Gladstone [1954] 1 W.L.R. 945; 98 S.J. 438; [1954] 2 All E.R. 206 39
Seward v. The Vera Cruz (1884) 10 App.Cas. 59, H.L.; *sub nom.* Vera Cruz, (The) (No. 2) (1884) 9 P.D. 96; 53 L.J.P. 33; 51 L.T. 104; 32 W.R. 783; 5 Asp.M.L.C. 270, C.A. 89

Shanley v. West Coast Stevedoring Co. [1957] 1 Lloyd's Rep. 391, C.A. 15
Sheppey v. Matthew T. Shaw & Co. [1952] W.N. 249; [1952] 1 T.L.R. 1272;
  96 S.J. 326 28
Shiels v. Cruikshank [1953] 1 W.L.R. 533; 97b S.J. 208; [1953] 1 All E.R. 874;
  sub nom. Cruikshank v. Shiels 1953 S.C. (H.L.) 1; 1953 S.L.T. 115,
  H.L.; affirming 1951 S.C. 741; 1952 S.L.T. 170 93
Shove v. Downs Surgical plc [1984] 1 All E.R. 7; [1984] I.R.C.R. 17 48
Simmons v. Boris [1956] 1 W.L.R. 381; 100 S.J. 283; [1956] 1 All E.R. 736 ...... 28, 29
Simpson v. Norwest Holst Southern [1980] 1 W.L.R. 968; 124 S.J. 313; 2 All
  E.R. 471, C.A. 52
Slater v. Hughes [1971] 3 All E.R. 1287, C.A. 50
Smith v. Austin Lifts [1959] 1 W.L.R. 100; 103 S.J. 73; [1959] 1 All E.R. 81,
  H.L.; [1958] 2 Lloyd's Rep. 583 .... 7, 10, 12, 19, 25
—— v. Baker & Sons [1891] A.C. 325, H.L.; 60 L.J.Q.B. 683; 65 L.T. 467; 7
  T.L.R. 679; 40 W.R. 392; 55 J.P. 660 .... 9, 37
—— v. Baveystock [1945] 1 All E.R. 531, C.A. 74
—— v. British Rail Engineering, *The Times*, June 27, 1980, C.A. 32
—— v. Chesterfield Co-operative Society [1953] 1 W.L.R. 370; 97 S.J. 132;
  [1953] 1 All E.R. 447, C.A.; 51 L.G.R. 194 69
—— v. Crossley Bros. 95 S.J. 655; (1951) C.L.C. 6831, C.A. 33
—— v. Leeds Brain & Co. [1962] 2 Q.B. 405; [1962] 2 W.L.R. 148; 106 S.J.
  77; [1961] 3 All E.R. 1159 .... 2, 3, 45
—— v. Ocean S.S. [1954] 2 Lloyd's Rep. 482 11
—— v. Scot Bowyers Ltd., *The Times*, April 16, 1986, C.A. 15
—— v. Chrysler (V.M.B.) (Scotland) Ltd. 1978 S.C. 1, H.L. 8
Sole v. Hallt (W.J.) Ltd. [1973] 1 Q.B. 574; [1973] 2 W.L.R. 171; [1973] 1 All
  E.R. 1032 81
Southern Water Authority v. Carey [1985] 2 All E.R. 1077 .... 8, 38
Stanley v. Concentric (Pressed Products) Ltd. (1972) 11 K.I.R. 260, C.A. 28
Stanton v. Ewart F. Youlden Ltd. [1960] 1 W.L.R. 543, 104; 104 S.J. 368;
  [1960] 1 All E.R. 429 92
Stanton Iron Works v. Skipper [1956] 1 Q.B. 255; [1955] 3 W.L.R. 752; 120
  J.P. 51; 99 S.J. 816; [1955] 3 All E.R. 544; 54 L.G.R. 25 63
Starr v. N.C.B. [1977] 1 All E.R. 243, C.A. 44
Staton v. N.C.B. [1957] 1 W.L.R. 893; [1957] 1 W.L.R. 893; 101 S.J. 592; 2 All
  E.R. 667 40
Staveley Iron & Chemical Co. v. Jones [1956] A.C. 627; [1956] 2 W.L.R. 479;
  100 S.J. 130; [1956] 1 All E.R. 403; *sub nom.* Jones v. Staveley Iron
  and Chemical Co. [1956] 1 Lloyd's Rep. 65 .... 33, 35
Steel v. Robert George & Co. Ltd. [1942] A.C. 497 46
Stewart v. West African Terminals Ltd. and J. Russell & Co. (1964) 108 S.J.
  838; [1964] 2 Lloyd's Rep. 371; affirming [1964] 1 Lloyd's Rep. 409 ...... 23, 27
Stitt v. Woolley (1971) 115 S.J. 708, C.A. 40
Stringer v. Automatic Woodturning [1956] 1 W.L.R. 138; [1956] 1 All E.R.
  327, C.A. 16
Summers (John) & Sons v. Frost [1955] A.C. 740; [1955] 2 W.L.R. 825; 99
  S.J. 257; [1955] 1 All E.R. 870; 53 L.G.R. 329 .... 35, 66, 68, 69
Sumner v. Priestley (R.L.) [1955] 1 W.L.R. 1202; 99 S.J. 780; [1955] 3 All
  E.R. 445; 54 L.G.R. 1; [1955] 2 Lloyd's Rep. 371 66

Taff Vale Ry. v. Jenkins [1913] A.C. 1; 57 S.J. 27; 82 L.J.K.B. 49; 107 L.T.
  564; 29 T.L.R. 19, H.L.; affecting S.C. *sub nom.* Jenkins v. Taff Vale
  Ry. (1912) 106 L.J. 715, C.A. 92
Taylor v. Ellerman's Wilson Line and Amos & Smith [1952] 1 Lloyd's Rep.
  144 28

xxiv Table of Cases

Tesco Stores Ltd. v. Edwards [1977] I.R.L.R. 120	115
Thomas v. British Aeroplane Co. [1954] 1 W.L.R. 694; 98 S.J. 302; [1954] 2 All E.R. 1, C.A.; 52 L.G.R. 292	28
Thompson v. Brown & Co. [1981] 1 W.L.R. 744; 125 S.J. 377; [1981] 2 All E.R. 296, H.L.	53
—— v. N.C.B. [1982] I.C.R. 15, C.A.	63
—— v. Price [1973] Q.B. 838; [1973] 2 W.L.R. 1037	93
—— v. Smiths Shiprepairers (North Shields) Ltd. [1984] 2 W.L.R. 522; 128 S.J. 255; [1984] 1 All E.R. 881; [1984] I.C.R. 236	32, 53
Thorne v. Strathclyde Regional Council (O.H.) 1984 S.L.T. 161	20
Thornton v. Swan Hunter (Shipbuilders) Ltd. [1972] 2 Lloyd's Rep. 112, C.A.	77
Thurogood v. Van Den Berghs and Jurgens [1951] 2 K.B. 537; [1951] 1 T.L.R. 557; 95 S.J. 317; 49 L.G.R. 504; sub nom. Thorogood v. Van Den Berghs and Jurgens, 115 J.P. 237; [1951] 1 All E.R. 682	12, 14
Tolley v. Morris [1979] 1 W.L.R. 592; 123 S.J. 353; [1979] 2 All E.R. 561, H.L.; affirming [1979] 1 W.L.R. 205; [1979] 1 All E.R. 71; (1978) 122 S.J. 437, C.A.	54
Tremain v. Pike [1969] 1 W.L.R. 1556; 113 S.J. 812; [1969] 3 All E.R. 1303	23, 45
Trim Joint District School Board v. Kelly [1914] A.C. 667, H.L.; 83 L.J.P.C. 220; 111 L.T. 305; 30 T.L.R. 452; 58 S.J. 493; 7 B.W.C.C. 274	104
Trott v. W.E. Smith [1957] 1 W.L.R. 1154; 101 S.J. 885; [1957] 3 All E.R. 500; 56 L.G.R. 20	71
Turner v. N.C.B. (1949) 65 T.L.R. 580, H.L.	22
Twine v. Bean's Express (1946) 175 L.T. 131; 62 T.L.R. 458, C.A.	39
Uddin v. Associated Portland Cement Manufacturers [1965] 2 Q.B. 582, C.A.; [1965] 2 W.L.R. 1183; 109 S.J. 313; [1965] 2 All E.R. 213; 63 L.G.R. 241, C.A.; affirming [1965] 2 Q.B. 15; [1965] 2 W.L.R. 327; 109 S.J. 151; [1965] 1 All E.R. 347	36
United Africa Co. v. Saka Owoade [1955] A.C. 130; [1955] 2 W.L.R. 13; 99 S.J. 26; [1957] 3 All E.R. 216; [1954] 2 Lloyd's Rep. 607	40
Upson v. Temple Engineering (Southend) [1975] K.I.L.R. 171	26
Vandyke v. Fender, Sun Insurance Office (Third Party) [1970] 2 Q.B. 292; [1970] 2 W.L.R. 929; 114 S.J. 205; [1970] 2 All E.R. 335; 8 K.I.R. 854	41
Vange Scaffolding and Engineering Co., The Times, March 26, 1987, C.A.	12
Vickers v. Gomme (E.) [1957] 1 W.L.R. 656; [1957] 1 W.L.R. 656; 101 S.J. 391; [1957] 2 All E.R. 60; 55 L.G.R. 222	71
Vincent v. P.L.A. [1957] 1 Lloyd's Rep. 103, C.A.	33
Vinnyey v. Star Paper Mills Ltd. [1965] 1 All E.R. 175	16
Virgo Steamship Co. SA v. Skaarup Shipping Corp. and Others, The Times, October 21, 1987	38
Voller v. Dairy Produce Packers Ltd. [1962] 3 All E.R. 938	94
Wagon Mound (The) [1961] A.C. 388, P.C.	44
Waite v. Redpath Dorman Long Ltd. [1971] 1 Q.B. 294; [1971] 1 All E.R. 5; [1970] 3 W.L.R. 1034	50
Walker v. Bletcheley Flettons [1937] 1 All E.R. 170	68
Walsh v. Allweather Mechanical Grouting Co. [1959] 2 Q.B. 300; [1959] 3 W.L.R. 1; 103 S.J. 489; [1959] 2 All E.R. 588; 52 R. & I.T. 402; 57 L.G.R. 153	31
—— v. Holst [1958] 1 W.L.R. 800; 102 S.J. 545; [1958] 3 All E.R. 33, C.A.	21, 22, 43

Walters v. Whessoe and Shell Refining Co. Ltd. (1960) 6 Build.L.R. 30; [1980] C.L.Y. 1520, C.A.	8
Warren v. Henlys [1948] W.N. 449; [1948] 2 All E.R. 935; 92 S.J. 706	40
—— v. Scruttons [1962] 1 Lloyd's Rep. 497	45
Warwick v. Jeffrey, *The Times*, June 21, 1983	92
Watson v. Ready Mixed Concrete, *The Times*, January 18, 1961	31
Watt v. Herts C.C. [1954] 1 W.L.R. 835; 118 J.P. 377; 98 S.J. 372; [1954] 2 All E.R. 368, C.A.; 52 L.G.R. 383	13
Watts v. Empire Transport Co., Ltd. [1963] 1 Lloyd's Rep. 263	18
Waugh v. British Railways Board [1979] 3 W.L.R. 150; 123 S.J. 506; [1979] 2 All E.R. 1169, H.L.; reversing (1978) 122 S.J. 730, C.A.	43
Welsford v. Lawford Asphalte Co. (May 16, 1956) unreported; [1956] C.L.Y. 5984	31
West Bromwich Building Society v. Townsend [1983] I.R.L.R. 147; [1983] I.C.R. 257, *The Times*, January 3, 1983, D.C.	11
West (H.) & Son v. Shephard [1964] A.C. 326; [1963] 2 W.L.R. 1359; 107 S.J. 474; [1963] 2 All E.R. 625, H.L.; affirming *sub nom*. Shephard v. West (H.) & Son (1962) 106 S.J. 817; affirming *sub nom*. Shephard v. West (H.) & Son (1962) 106 S.J. 391	46, 47
Westwood v. Post Office [1974] A.C. 1, H.L.; [1973] 3 W.L.R. 287; [1973] 3 All E.R. 184, H.L.	63
Whalley v. Briggs Motors [1954] 1 W.L.R. 840; 98 S.J. 373; [1954] 2 All E.R. 193	63
Wheeler v. London and Rochester Trading Co. [1957] 1 Lloyd's Rep. 69	15
Whincup v. Woodhead (Joseph) & Sons (Engineers), 115 J.P. 97; [1951] 1 All E.R. 387	64
Whitby v. Burt Boulton & Hayward [1947] K.B. 918; [1947] K.B. 918; [1947] L.J.R. 1280; 177 L.T. 556; 63 T.L.R. 458; 111 J.P. 481; 91 S.J. 517; [1947] 2 All E.R. 325	64
White v. Holbrook Precision Castings [1985] I.R.L.R. 215, C.A.	16
Whitehead v. Stott (James) [1949] 1 K.B. 358; [1949] L.J.R. 1144; 65 T.L.R. 94; 93 S.J. 58; 1 All E.R. 245, C.A.; 47 L.G.R. 222	75
Wreland v. Cyril Lord Carpets [1969] 3 All E.R. 1006	48
Wigley v. British Vinegars [1964] A.C. 307; [1963] 3 W.L.R. 731; 106 S.J. 609; [1962] 3 All E.R. 161, H.L.; affirming [1961] 1 W.L.R. 1261	69, 73
Wilkins v. William Cory & Son [1959] 2 Lloyd's Rep. 98	48
Wilkinson v. Ancliff (BLT) Ltd. [1986] 1 W.L.R. 1352 at 1365; [1986] 3 All E.R. 427 at 438, C.A.	53
Willard v. Whiteley 82 S.J. 711; [1938] 3 All E.R. 779, C.A.	42
Wilsher v. Essex Area Health Authority *The Times*, March 11, 1988, H.L.	22
Wilson v. International Combustion (July 18, 1956) C.A. noted in [1957] 3 All E.R. 505	15
—— v. Tyneside Window Cleaning [1958] 2 Q.B. 110; [1958] 2 W.L.R. 900; 102 S.J. 380; [1958] 2 All E.R. 265	15, 19, 26
Wilsons and Clyde Coal Co. v. English [1938] A.C. 57, H.L.; 106 L.J.P.C. 117; 157 L.T. 406; 53 T.L.R. 944; 81 S.J. 700; [1937] 2 All E.R. 628	9
Wingfield v. Ellerman's Wilson Line (Furley & Co., Third Parties) [1960] 2 Lloyd's Rep. 16, C.A.	10
Winter v. Cardiff R.D.C. [1950] W.N. 193, H.L.; 114 J.P. 234; [1950] 1 All E.R. 819; 49 L.G.R. 1	24
Wise v. Kaye [1962] 1 Q.B. 638, C.A.; [1962] 2 W.L.R. 96; 106 S.J. 14; [1962] 1 All E.R. 257, C.A.	46
Withers v. Perry Chain Co. [1961] 1 W.L.R. 1314, C.A.; 105 S.J. 648; [1961] 3 All E.R. 676; 59 L.G.R. 496	9, 14
Wood v. Dutton's Brewery Ltd. (1971) 115 S.J. 186, C.A.	33

Woods v. Duncan; Duncan v. Cammell Laird & Co. [1946] A.C. 401; [1946] 1
All E.R. 420; [1947] L.J.R. 120; 174 L.J. 286; 62 T.L.R. 283 22
—— v. Durable Suites [1953] 1 W.L.R. 857; 97 S.J. 454; [1953] 2 All E.R.
391, C.A.; 51 L.G.R. 424 .... 15, 16, 31
Woodward v. Renold Ltd. [1980] I.C.R. 387 28
Woollins v. British Celanese (1966) 110 S.J. 686 82
Wright v. British Railways Board [1983] 3 W.L.R. 211; [1983] 2 All E.R. 698,
H.L. 51

Yorke v. British and Continental S.S. Co. (1945) 78 Ll.L.Rep. 181, C.A. 65
Young (John) & Co. (Kelvinhaugh) Ltd. v. O'Donnell, 1958 S.L.T. (Notes)
46 .... 41, 42
Youngman v. Pinelli General Cable Works [1940] 1 K.B. 1; 109 L.J.K.B. 420,
C.A.; 160 L.T. 489 20

Table of Statutes

1802	Factory Act (42 Geo. 3, c. 73)	2	1945	Law Reform (Contributory Negligence) Act (8 & 9 Geo. 6, c. 28)—
1833	Factory Act (3 & 4 Will. 4, c. 103)— ss. 2–9	110		s. 1(1) 34
1846	Fatal Accidents Act (9 & 10 Vict. c. 93) .... s. 1	89 89, 90	1947	Crown Proceedings Act (10 & 11 Geo. 6, c. 44) 65, 85
1894	Merchant Shipping Act (57 & 58 Vict. c. 60)— s. 503	50		s. 2(1)(b) .... 7 (2) .... 65 (3) .... 65 s. 6 .... 84 s. 10 .... 8, 65
1930	Reservoirs (Safety Provisions) Act (20 & 21 Geo. 5, c. 51)	62	1948	Law Reform (Personal Injuries) Act (11 & 12 Geo. 6, c. 41) ... 49, 102
1934	Law Reform (Miscellaneous Provisions) Act (24 & 25 Geo. 5, c. 41) .... 86, 87, 89, 91, 95 s. 1 .... 86 (2)(c) .... 87, 88 (5) .... 87, 95 s. 3 50			s. 2 .... 102 (1) 49 (2) 49
			1954	Mines & Quarries Act (2 & 3 Eliz. 2, c. 50) 2, 61, 110, 113, 131
			1956	Agriculture (Safety, Health & Walfare Provisions) Act (4 & 5 Eliz. 2, c. 49) ...... 2, 3, 61 s. 7 .... 3 s. 22 .... 65 s. 24(1) 2
1935	Law Reform (Married Women and Tortfeasors) Act (25 & 26 Geo. 5, c. 30)— s. 6	38	1957	Occupiers' Liability Act (5 & 6 Eliz. 2, c. 31) .... 7, 79, 82, 83 s. 1(1) .... 52, 79 (c) .... 82 (2) .... 79, 80 (3)(a) .... 80 (b) .... 80 s. 2(1) .... 82 (3) .... 82 (a) .... 82 (b) .... 82 (4) .... 83 (a) 83
1936	Public Health Act (26 Geo. 5 & 1 Edw. 8, c.49)— s. 45	136		
	Public Health (London) Act (26 Geo. 5 & 1 Edw. 8, c. 50)— s. 106	136		
1937	Factory Act (1 Edw. 8 & 1 Geo. 6, c. 67) .... s. 4	2 67		
1939	Limitation Act (2 & 3 Geo. 6, c. 21) .... s. 21	52 52		

xxvii

xxviii Table of Statutes

1957	Occupiers' Liability Act —*cont.*	
	s. 2 (4)—*cont.*	
	(*b*)	84
	(6)	81
	s. 3 (1)	82
	s. 4	80
	s. 5 (1)	81
	(2)	81
	(3)	81
	s. 25	83
1958	Merchant Shipping (Liability of Shipowners & Others) Act (6 & 7 Eliz. 2, c. 62)—	
	s. 3 (2) (*a*)	51
1961	Factory Act (9 & 10 Eliz. 2, c. 31)	2, 18, 61, 79, 110, 113, 114, 115
	s. 1	**116**
	s. 2	**117**
	s. 3	**118**
	s. 4	66, **119**
	s. 5	**119**
	s. 6	**119**
	s. 7	**119**
	s. 10A	**120**
	s. 11	**121**
	s. 12	**122**
	s. 13	**122**
	s. 14	2, **123**
	(1)	77
	s. 15	**123**
	s. 16	**124**
	s. 17 (2)	64
	s. 28	72, **124**
	(1)	70
	s. 29	28, **124**
	(1)	70
	s. 57	**125**
	s. 58	**125**
	s. 59	**126**
	s. 60	**126**
	s. 74	97
	s. 123 (1)	132
	s. 125 (1)	131
	s. 173	65
	s. 175	2, **126**
1963	Offices, Shops & Railway Premises Act (c. 41)	2, 3, 61, 79, 110, 113, 114, 115, **129**
	s. 1	2, **130**
	s. 2	2, **132**
	s. 3	2, **132**
	s. 4	**133**
1963	Offices, Shops & Railway Premises Act—*cont.*	
	s. 5	**133**
	s. 6	3, **134**
	s. 7	3, **135**
	s. 8	3, **136**
	s. 9	**136**
	s. 10	**137**
	s. 11	**137**
	s. 12	10, **138**
	s. 13	**139**
	s. 14	**139**
	s. 15	**140**
	s. 16	**140**
	s. 17	**141**
	s. 18	**141**
	s. 83	65
1964	Industrial Training Act (c. 16)	96
1965	Nuclear Installations Act (c. 57)	62
	s. 12	150
	s. 47 (1)	62
	(2)	62
1969	Employer's Liability (Defective Equipment) Act (c. 37)	11, 30
	s. 1 (1)	30
	(2)	51
1969	Enforcement of the Employers' Liability (Compulsory Insurance) Act (c. 57)	113
1970	Income & Corporation Taxes Act (c. 10)—	
	s. 187	48
	Administration of Justice Act (c. 31)	43
	s. 6	44
	ss. 31–35	43, 46
1970	Merchant Shipping Act (c. 36)—	
	ss. 19–26	61
	Chronically Sick & Disabled Persons Act (c. 44)	85
	s. 8A	85
1971	Fire Precautions Act (c. 40)	113
	s. 41 (1)	113
	Mineral Workings (Offshore) Installations) Act (c. 61)	62
1972	Gas Act (c. 60)	62
	European Communities Act (c. 68)	3

Table of Statutes **xxix**

1974	Health and Safety at Work etc. Act (c. 37) 1, 62, 65, 79, 100, 111, 114, 115	
	s. 1 .... **142**	
	ss. 2–9 110	
	s. 2 .... **143**	
	(4) 98, 99	
	(6) 98	
	s. 3 .... **144**	
	s. 4 .... **145**	
	s. 5 .... **145**	
	s. 6 .... **146**	
	s. 7 .... **150**	
	s. 8 .... **150**	
	s. 11 111	
	(4) 111	
	s. 12 111	
	s. 13 113	
	(1)(d) 111	
	s. 14 111	
	s. 15(6)(b) 150	
	s. 16 .... 58, 99, 111	
	s. 17 .... 3, 111, 113	
	ss. 18–26 3	
	s. 19 113	
	ss. 20–26 114	
	s. 20 114	
	s. 21 115	
	s. 22 115	
	s. 24 115	
	s. 25 114	
	s. 27 114	
	s. 28 114	
	s. 33 115	
	s. 34 115	
	s. 35 115	
	s. 36 115	
	s. 37 115	
	s. 38 115	
	s. 39 115	
	s. 40 115	
	s. 41 115	
	s. 42 115	
	s. 47 .... 115, **150**	
	s. 48 65	
	s. 78 113	
	s. 80 114	
	Sched. 1 .... 110, 114	
	Pt. I 3	
	Pt. II 3	
	Pt. IV 3	
1975	Social Security Act (c. 14) 6, 107	
	ss. 1–13 105	
1975	Social Security Act—cont.	
	s. 17(1) 102, 105	
	s. 36 .... 102, 105	
	ss. 50–78 103	
	ss. 50 102	
	s. 50A 104	
	s. 50(1) 103	
	s. 51 .... 102, 103	
	s. 52 .... 102, 103	
	s. 53 .... 40, 102, 103, 104	
	s. 54 .... 102, 103	
	s. 55 .... 102, 103	
	s. 57 106	
	s. 76 101	
	s. 77 101	
	s. 78 101	
	s. 94 109	
	s. 97 107	
	(3) 108	
	s. 98 107	
	s. 100 107	
	s. 101 108	
	ss. 108–111 109	
	s. 112 109	
	s. 116 108	
	s. 156 .... 101, 103	
	Sched. 10 107	
	para. 1(8) 108	
	Chap. IV 104	
1975	Reservoirs Act (c. 23) 62	
	Limitation Act (c. 54) 91	
	Petroleum and Submarine Pipe-lines Act (c. 74) 62	
1976	Fatal Accidents Act (c. 30) 86, 87, 89, 91	
	s. 1 89	
	(3) 90	
	(4) 90	
	(5) 90	
	s. 1A 94	
	(2) 94	
	(4) 94	
	s. 2 89	
	s. 3 92	
	(1) 89	
	(3) 93	
	s. 4 92	
	s. 5 91	
	s. 34 94	
	Chronically Sick and Disabled Persons (Amendment) Act (c. 49) 85	

1977	Unfair Contract Terms Act (c. 50)	78, 82, 83	1979	Pneumoconiosis, etc., Act—*cont.*	
	ss. 1–14	82		s. 4—*cont.*	
	s. 1	50		(2)	109
	(3)	83		s. 5	109
	s. 2	37, 50	1980	Limitation Act (c. 58)	52
	(3)	83		s. 11	52, 53, 91
	s. 11	52, 82		s. 11 (5)	53
	s. 14	50, 51		s. 12 (1)	91
	s. 16 (3)	37		s. 13	91
	para. 4	52		(1)	53
	Sched. 1, para. 5	52		s. 14	52
1978	State Immunity Act (c. 33)—			s. 33	53, 91
	s. 4	8, 65		s. 38 (1)	52
	s. 5	8, 65	1982	Social Security Act (c. 2)	104
	s. 14	65		Industrial Training Act (c. 10)	96
	Employment Protection (Consolidation) Act (c. 44)—			Oil and Gas (Enterprise) Act (c. 23)	62
	s. 19	58		Social Security and Housing Benefits Act (c. 24)	49, 57
	(1)	57, 58		s. 1 (2)	57
	s. 20	58		(3)	57
	(1)	59		s. 5 (4)	57
	(2)	59		s. 7 (1)	57
	(4)	59		s. 9	58
	s. 21	58, 59		Employment Act (c. 46)	59
	s. 22	58, 60			
	ss. 54–80	97		Administration of Justice Act (c. 53)	50, 95
	s. 55 (2) (c)	98		s. 1	47
	s. 57	97, 98		(1)	46
	(2) (d)	97		(a)	88
	(3)	97		(2)	45
	s. 64 (1) (a)	59		s. 3	89, 90, 92, 93, 94, 95
	(2)	59, 97		(1)	88, 94
	Sched. 1	58		s. 4 (1)	94
	Civil Liability (Contribution) Act (c. 47)	11		(2)	88
	s. 1	38		s. 5	46
	s. 2	38		s. 6	48
	s. 3	38		s. 15	50
1979	Merchant Shipping Act (c. 39)	61		Sched. 1	50
	s. 21	61	1984	Occupiers' Liability Act (c. 3)	79, 81, 84, 85
	s. 22	61		s. 1 (a)	84
	s. 23	61		(3)	85
1979	Pneumoconiosis, etc., (Workers' Compensation) Act (c. 41)	101, 105, 106		(5)	85
				(6)	85
	s. 1 (3)	106		(7)	84
	s. 2 (1) (b)	105		(9)	84
	(4)	106		s. 2	83
	s. 3	106		s. 3	85
	s. 4	109			

1984	Health and Social Security Act (c. 48)—	
	s. 11	105
1986	Safety at Sea Act (c. 23)	61
	Social Security Act (c. 50)	104
	Sched. 3	104
	Sched. 5, para. 7	108
	Sex Discrimination Act (c. 59)	97
1986	National Health Service (Amendment) Act (c. 66)	7, 65
1987	Crown Proceedings (Armed Forces) Act (c. 25)	65
	Fire Safety and Safety of Places of Sport Act (c. 27)	113
	Crown Proceedings (Armed Forces) Act (c. 25)	8
	s. 1	65
	s. 2	65

Table of Statutory Instruments

Year	Instrument	Page
1922	India Rubber Regulations S.R. and O. (S.I. 1922 No. 329)— reg. 12	58
	Chemical Works Regulations S.R. and O. (S.I. 1922 No. 731)— reg. 30	58
1934	Docks Regulations (S.I. 1934 No. 279)	65
1938	Operations at Unfenced Machinery Regulations (S.I. 1938 No. 641)	77
1948	Buildings Regulations (S.I. 1948 No. 1145)	63
1959	Quarries (Explosives) Regulations (S.I. 1959 No. 2259)	77
1966	Construction (Working Places) Regulations (S.I. 1966 No. 94)	42
1968	Ionising Radiations (Unsealed Radioactive Substances) Regulations (S.I. 1968 No. 780)— regs. 12, 33	58
1969	Ionising Radiations (Sealed Sources) Regulations (S.I. 1969 No. 808)— regs. 11, 30	58
1970	Radioactive Substances (Road Transport Workers) (Great Britain) Regulations 1970 (as amended by S.I. 1975 No. 1522) (S.I. 1970 No. 5973)— reg. 14	59
1971	Employers' Liability (Compulsory Insurance) Exemption Regulations (S.I. 1971 No. 1933)	55
1974	Excise Warehousing Regulations (S.I. 1974 No. 208)	55
	Health and Safety at Work etc. Act (Application outside Great Britain) Order (S.I. 1977 No. 1232)	66
	Protection of Eyes Regulations (S.I. 1974 No. 1681)	31
	Factories Act 1961 etc. (Repeals and Modifications) Regulations (S.I. 1974 No. 1941)	117, 118, 119, 120, 123, 125, 126
	Offices, Shops and Railway Premises Act 1963 (Repeals and Modifications) Regulations (S.I. 1974 No. 1943)	133, 135, 136, 137, 139
1975	Employers' Liability (Compulsory Insurance) (Amendment) Regulations (S.I. 1975 No. 194)	113
	Social Security (Employed Earners' Employments for Industrial Injuries Purposes) Regulations (S.I. 1975 No. 467)	103

1975 Offshore Installations (Application of the Employers' Liability (Compulsory Insurance) Act 1969) Regulations (S.I. 1975 No. 1289)	55
Employers' Liability (Compulsory Insurance) (Offshore Installations) Regulations (S.I. 1975 No. 1443)	55
1976 Fire Certificates (Special Premises) Regulations (S.I. 1976 No. 2003)	114
Fire Precautions (Factories, Offices, Shops and Railway Premises) Order (S.I. 1976 No. 2009)	113
Fire Precautions (Non-Certificated Factory, Office, Shop and Railway Premises) Regulations (S.I. 1976 No. 2010)	113
1977 Safety Representatives and Safety Committees Regulations (S.I. 1977 No. 500)	98
reg. 4	99
(2)	99
reg. 5	99
reg. 11	99
(3)	99
Sched.	99, 100
Health and Safety (Enforcing Authority) Regulations (S.I. 1977 No. 746)	113
1978 State Immunity Act 1978 (Commencement) Order (S.I. 1978 No. 1572)	8
1980 Control of Lead at Work Regulations (S.I. 1980 No. 1248)—	
reg. 16	59
1981 Employers' Liability (Compulsory Insurance) (Amendment) Regulations (S.I. 1981 No. 1489)	55
1982 Statutory Sick Pay (General Regulations (S.I. 1982 No. 894)	57
Social Security Benefit (General Benefit) Regulations (S.I. 1982 No. 1408)	104, 105
1984 Contract of Industrial Major Hazards Regulations (S.I. 1984 No. 1902)	111
1985 Social Security (Industrial Injuries) (Prescribed Diseases) Regulations (S.I. 1985 No. 967)	103, 106
reg. 2	103
(c)	103
reg. 4	103
(1)	103
Ionising Radiations Regulations (S.I. 1985 No. 1333)	111
Pneumoconiosis, etc., (Workers' Compensation) (Payment of Claims) Regulations (S.I. 1985 No. 2035)	109
1986 Social Security (Adjudication) Regulations (S.I. 1986 No. 2218)	107
reg. 24(2)	108
regs. 27–36	109
1987 Social Security Commissioners Procedure Rules (S.I. 1987 No. 214)	107
1988 Pneumoconiosis, etc., (Workers' Compensation) (Payment of claims) Regulations (S.I. 1988 No. 668)	106

Rules of the Supreme Court

Ord. 29, r. 9	44

Table of Abbreviations

Law Reports
A.C.	Appeal Cases (Law Reports)
All E.R.	All England Law Reports
C.L.	Current Law
C.L.Y.	Current Law Yearbook
Ch.	Chancery (Law Reports)
I.C.R.	Industrial Cases Reports
I.R.L.R.	Industrial Relations Law Reports
K.I.R.	Knight's Industrial Reports (Law Reports)
Lloyd's Rep.	Lloyd's Reports (Law Reports)
Q.B.	Queen's Bench (Law Reports)
S.C.	Session Cases (Scottish Law Reports)
S.J.	Solicitors' Journal
S.L.T.	Scottish Law Times
T.L.R.	Times Law Reports
W.N.	Weekly Notes (Law Reports)
W.L.R.	Weekly Law Reports

Courts
C.A.	Court of Appeal
D.C.	Divisional Court
E.A.T.	Employment Appeal Tribunal
H. Ct.	High Court
H.L.	House of Lords
I.T.	Industrial Tribunal
Ct. Sess.	Court of Session (Scotland)

1. Introduction and Sources

(1) INTRODUCTORY

This book is written for those, lawyers and laymen, who wish for an overall, though reasonably detailed, view of the law and practice as it affects health and safety at work and in particular the relationship between employers and employees in regard to health and safety of the employees. "Employee" here may well have a wider meaning than just that in which Victorian lawyers used the word "servant." Much of the legislative protection may well apply also to those on employer's premises, whether employees in the narrow sense or not, who come in contact with those premises or with machinery, etc., on them. Mere visitors, or those who come to render services for a fee, may well also be within the ambit of the legislation. This book does not however deal with the relationship of employers with the general public outside, concerning for example the problem of whether or not noxious emissions from factories, etc., can create rights of action in the general public. Public safety from such emissions, explosions, etc., has been the subject of increasing legislation in recent years. The tendency of some of the legislation has been to blur the distinction between the protection of those working on the premises and the protection of members of the general public outside them but the scope of this book can include only the former. It is not uncommon nowadays, for example, for regulations which are intended to protect the general public to be made under the Health and Safety at Work, etc., Act 1974 or under public health or consumer protection legislation. Whatever its source, such legislation cannot be the subject of this book, since the subject of health and safety at work proper is a very large one and the legislation is voluminous but this book endeavours to indicate the content of the principal legislation (Acts of Parliament and Regulations) which relate to health and safety at work. This book is written from the standpoint of individuals who may be affected by the law rather than the view of the enforcing authorities such as Local Authorities or the Health and Safety Executive, though Chapter 5 of the book does deal in some

2 Introduction and Sources

detail with the powers of the Health and Safety Executive and other enforcing authorities.

(2) SOURCE MATERIALS

(a) Acts of Parliament and Regulations made under those Acts

Since the Factory Acts of the nineteenth Century, health and safety at work has been the subject of increasing legislative regulation. The first Factories Act was passed in 1802 and was simply, "an Act for the preservation of the health and morals of apprentices and others employed in cotton and other mills and cotton and other factories" but gradually the legislation became more comprehensive culminating (with the important landmark of the Factories Act 1937) in the Factories Act 1961. The 1961 Act, together with the regulations made under it and numerous regulations made under earlier Factories Acts still constitutes much of the law on this subject. The word "factory" is given an extremely wide definition[1] so that not only the conventional notion of a factory is included but also ancillary premises such as laundries, premises for the sorting of articles, etc. Nevertheless the definition of "factory" is not so comprehensive as to include by any means all places in which persons work. "Offices, Shops and Railway Premises" constituted a considerable new extension of legislative regulation by the 1963 Act of that title.[2] In other spheres, the Agriculture (Safety, Health and Welfare Provisions) Act 1956 (and regulations made thereunder) introduced legislative regulation for an entirely new area, that of "agriculture."[3] Moreover, for many years mines and quarries have been the subject of legislative regulation, the principal Act being the Mines and Quarries Act 1954. The subject of mines and quarries is very much one of its own and is the subject of very detailed and specialised factual regulation. By and large therefore it will not be dealt with in this book.

All the above-cited Acts and a very large number of regulations made under them constitute detailed legal and factual regulation in specific circumstances within the areas covered, such as for example the duty securely to guard dangerous machinery under section 14 of the Factories Act 1961, the duty to provide adequate

---

[1] By s.175 of the 1961 Act.
[2] For the detailed definition of offices, shops and railway premises see sections 1–3 of the 1963 Act.
[3] Defined in s.24(1) of the 1956 Act, in a wide manner.

ventilation and lighting in the Offices, Shops and Railway Premises Act 1963[4] and the regulation-making power in the Agriculture (Safety, Health and Welfare Provisions) Act 1956 to prohibit children from riding on or driving vehicles, machinery or implements used in agriculture.[5]

In 1974 radical new legislation was enacted in the form of the Health and Safety at Work, etc., Act 1974, Parts I, II and IV of which[6] implemented the basic recommendations of the Report of the Robens Committee on Safety and Health at Work.[7] That Report recommended the creation of a more unified and integrated system of legislation on health and safety at work which would progressively by regulations made under the 1974 Act replace the existing statutory provisions listed. But although many regulations have now been made under the Health and Safety at Work, etc., Act 1974, the corpus of the other Acts and regulations made under them is still extremely important and far reaching. The 1974 Act imposed a series of general duties on employers owed to their employees, to those who were self-employed and to others entering upon employers' premises[8] and provided for the making of "Health and Safety Regulations" with attendant Codes of Practice and Guidance Notes formulated by the then newly created Health and Safety Commission. Many such have been promulgated since 1974. The Codes of Practice and Guidance Notes do not have the same force of law as the Regulations themselves but are of course nevertheless strong evidence in criminal or civil proceedings.[9] The 1974 Act also created a new Health and Safety Commission and Health and Safety Executive, the Commission's enforcement arm,[10] and gave the Health and Safety Executive stringent enforcement powers.[11] Regulations that have been made since the coming into operation of the 1974 Act have not only been made that Act but also under the provisions of the European Communities Act 1972, requiring the United Kingdom to implement EEC Directives and these are forming an increasingly important part of Health and Safety Law. A selection of the statutory materials (Acts and Regulations, etc.) is to be found in Appendices 1 and 2 of this work but of necessity in a short work, only selected sec-

---

[4] ss.6–8.
[5] 1956 Act, s.7.
[6] Pt. III is concerned with building regulations, etc., and is not relevant to this work.
[7] 1970–1972 Cmnd. 5034.
[8] 1974 Act, ss.2–9.
[9] Compare s.17 of the 1974 Act.
[10] See also Chap. 5 below.
[11] See ss.18–26 of the 1974 Act and see Chap. 5 below.

tions, etc., can be reproduced and the reader who requires access to all of the Acts and Regulations should for example consult the "Encyclopedia of Health and Safety at Work."[12]

In addition to the statutory materials a very large number of publications on a multiplicity of topics are issued and kept up-to-date by the Health and Safety Commission and the Health and Safety Executive. Although of course those publications do not constitute the law as such, they form a formidable corpus of quasi-law. For a complete list of them the reader should consult the Encyclopedia of Health and Safety at Work. Compliance with Guidance Notes and other literature issued by the Health and Safety Commission and the Health and Safety Executive would of course be relevant in deciding for example whether, if there had been an accident,[13] an employer had followed the best current practice and was not therefore negligent.

---

[12] Sweet and Maxwell, three volumes, a loose leaf work kept regularly up-to-date.
[13] Compare Chap. 2 below.

2. Claims by Employees and Others in the Courts

(1) CLAIMS FOR DAMAGES FOR DISEASES AND PERSONAL INJURIES (AT COMMON LAW AND FOR BREACH OF STATUTORY DUTY)

(a) Introductory

An employee who is injured or who contracts a disease at work, or elsewhere where he is working, may be able to claim damages from his employer in the courts.[1] Moreover, the statutory duties imposed upon an employer[2] may also extend to persons other than employees, *e.g.* visitors and those who are self-employed and rendering services to the employer for a fee.[3] Such damages are not available just because the injury or the disease was caused by the work situation[4] but are recoverable only if the employee, etc., can establish a cause of action (a recognised head of claim) within the general structure of the civil law.[5]

(b) Claims at common law

Nature of liability

A claim at "common law" means a claim which can be maintained under the law as established by decided cases over the centuries, though much of the law on the present subject is in comparatively recent case law, *i.e.* largely during the years of the present century.

---

[1] For claims in industrial tribunals and for social security benefits see Chaps. 3 and 4 below.
[2] See pp. 61 *et seq.*
[3] See pp. 62–64.
[4] Compare social security benefits pp. 101 *et seq.*
[5] Criminal breaches of statute may not necessarily ground civil liability in damages. See p. 62.

A common-law claim is to be distinguished from a claim for breach of statutory duty[6] (*i.e.* duty imposed by an Act or Regulation) or a claim for social security benefits under the Industrial Injuries and Diseases Scheme now embodied in the Social Security Act 1975 and Regulations made under that Act.[7] So far as common law is concerned, the case law establishes that in practice the only "tort" (*i.e.* civil wrong remediable in damages) that is in practice applicable to claims by employees, etc., is the tort of Negligence.[8] This relates of course to English Law. The law may be different abroad, *e.g.* the case may be able to be brought only as an action for a breach of contract.[9] Moreover, it may be possible to secure much higher damages in another country, in which case problems may arise where it is sought to restrain an employee from suing in such other country.[10]

Constituents of tort of negligence

These constituents are the rules applicable to the tort generally but they also have application to the type of claim we are now considering. It is necessary in order to establish a claim for negligence that the plaintiff employee must show:

(1) that the employer owed him a duty of care;
(2) that the employer was in breach of that duty; and
(3) that as a result of that breach the employee suffered damage.

If those elements can be established, then the employee's claim will succeed, unless there are any special defences open which would defeat the claim either in whole or in part.[11] These elements are discussed in detail below.

Who can be the Plaintiff (i.e. the person making the claim)

The plaintiff will normally be an employee, *i.e.* a person who works under a contract of service with an employer by which the employer is in theory at least able to direct the employee not only as to what work he does but the manner in which he does it.[12] If

---

[6] As to which see pp. 61 *et seq.*
[7] See pp. 101 *et seq.*
[8] In theory Nuisance and Trespass to the Person are also available. See *Letang* v. *Cooper* [1965] 1 Q.B. 232, C.A.; *Barnes* v. *Nayer*, *The Times*, December 19, 1986, C.A.
[9] *Matthews* v. *Kuwait Bechtel Corporation* [1959] 2 Q.B. 57, C.A.
[10] See *MacShannon* v. *Rockwear Glass Ltd.* [1978] 1 All E.R. 625, H.L.; *Castanho* v. *Brown and Root (U.K.) Ltd. and Another* [1980] 3 W.L.R. 991, H.L.; *Coupland* v. *Arabian Gulf Oil Company* [1983] 3 All E.R. 226.
[11] See pp. 34 *et seq.*
[12] See further vicarious liability, pp. 37–43.

there is the true employer/employee relationship thus established (known formerly as the relationship of master and servant) then the duties of care owed by the employer will very conceivably be higher than those which would be owed to a person who was not an employee but was merely rendering services for a fee, *i.e.* a self-employed contractor.[13] It should be borne in mind that the court will, however, not necessarily accept a 'label' describing the workman as not an employee since this would otherwise be a fruitful method of evasion by employers of their duties. A court may well hold that a person described as a self-employed independent contractor is in fact an employee under a contract of service, *i.e.* a servant.[14]

Who can be a Defendant (the person against whom the claim is made)

The usual defendant is of course the employer but it may well be that the employee is also able to sue, *e.g.* the manufacturer of defective equipment supplied to the employer[15] or the occupier of premises to whom the employee is sent, if the employee is injured by the condition of those premises.[16] Examples of particular defendants are given below.

The Crown

Where the employer is the Crown[17] it is basically under the same liability to its employees as any other employer.[18] Formerly the Crown was not liable for injury sustained by members of the

---

[13] *Quinn* v. *Burch Bros. (Builders) Ltd.* [1966] 2 Q.B. 370, C.A.; *Jones* v. *Minton Construction Ltd.* (1974) 15 K.I.R. 309; compare the prison cases, *Ferguson* v. *Home Office*, The Times, October 8, 1977; *Pullin* v. *Prison Commissioners* [1957] 1 W.L.R. 1186.

[14] *Ferguson* v. *John Dawson and Partners (Contractors)* [1976] 1 W.L.R. 1213, C.A. (builders' labourers in fact employees); *Nethergate Ltd.* v. *Taverna*, The Times, May 4, 1984, C.A. (homeworker an employee).

[15] See on this the provisions of the Employers Liability (Defective Equipment) Act 1969, p. 11.

[16] See Occupiers' Liability Act 1957, and compare *Smith* v. *Austin Lifts* [1959] 1 W.L.R. 100.

[17] *i.e.* not the Sovereign personally, but includes, *e.g.* Government Departments, HM Forces, and a number of other bodies such as the Forestry Commission, Medical Research Council. Nationalised industries are not Crown bodies, neither in relation to Health and Safety Legislation is the National Health Service—National Health Service (Amendment) Act 1986.

[18] Crown Proceedings Act 1947, s.2(1)(*b*).

Armed Forces[19] but by the Crown Proceedings (Armed Forces) Act 1987 henceforth the Crown is to be liable for such injuries if caused, *e.g.* by negligence but there are qualifications, both as to time and other matters.[20]

Foreign States

Foreign States are equally liable to their employees for deaths or personal injury to those employees caused by acts or omissions in the United Kingdom, but only of course to the extent that a United Kingdom employer would be liable.[21]

Claims for Contribution or Indemnity

Any defendant (*e.g.* employer or occupier) who is made liable to an employee for damages will have a claim for a contribution of a proportion of the damages he has to pay, or an indemnity for complete reimbursement of the damages and costs, against any other person or body which by its negligence contributed to or caused the injury or disease to the employee.[22] This right to contribution or indemnity will arise according to the proportions of the blame but in addition an employer may also by a contract with a third party (*e.g.* on whose premises the employer is carrying out work) have stipulated for a right of contribution or indemnity from the third party, though such a contract will not be interpreted as indemnifying the employer against his own negligence except where its wording is capable of no other construction.[23]

Where there has been a loan of an employee by one employer to another the question may arise whether the employee can sue one or both of the employers, *i.e.* the lender and the borrower respect-

---

[19] Though a civilian hospital to which an injured Serviceman is taken may be if it is negligent—*Bell* v. *Secretary of State for Defence* [1985] 3 All E.R. 661, C.A.; *Pearce* v. *Secretary of State for Defence*, *The Times*, December 31, 1986.

[20] See Crown Proceedings Act 1947, s.10 and the Crown Proceedings (Amendment) Act 1987.

[21] State Immunity Act 1978, ss.4 and 5 (brought into force on November 22, 1978 by S.I. 1978 No. 1572).

[22] This is now governed by the Civil Liability Contribution Act 1978. As to contribution between independent tortfeasors see *Fitzgerald* v. *Lane*, *The Times*, March 7, 1987, C.A.

[23] See *Morris* v. *Ford Motor Company Ltd.* [1973] 1 Q.B. 792, C.A.; *Walters* v. *Whesso and Shell Refining Co. Ltd.* (1960) 6 Build.L.R. 30; [1980] C.L.Y. 1520, C.A.; *Smith* v. *U.M.B. Chrysler (Scotland) Ltd.* 1978 S.C.1 H.L.; *R. H. Green and S. W. Ltd.* v. *British Railways Board*, *The Times*, October 8, 1980; *Harper* v. *Grey-Walker* [1985] 2 All E.R. 507; *Southern Water Authority* v. *Carey* [1985] 2 All E.R. 1077; *Ronex Properties Ltd.* v. *John Laing Ltd.*, *The Times*, July 28, 1982, C.A.

ively, or both of them.[24] This matter is more fully dealt with under the head of vicarious liability[25] but the test is the same, namely whether or not the borrowing employer has the right of control of the servant so as to be liable to him as an employer. Even in such a case the original employer may well be liable but it is impossible to formulate a general rule since it depends on the facts of the case.[26]

(2) DUTY OF CARE

General scope of duty

The general principle is that the employer is bound to take *reasonable* care for the safety of his employees. The duty is often subdivided into a duty of taking reasonable care to provide proper appliances, machinery and plant, to maintain them in a proper condition, and to carry on that operation so as not to subject the employees to unnecessary risk. Although it is convenient to consider the duty of care under those heads they are but examples of the general principle to take reasonable care.[27]

Special susceptibility of employee

As the duty is only to take reasonable care, which is an objective test, the question arises what is the position if the employee is abnormally susceptible, *e.g.* to the contraction of industrial diseases. The position would appear to be that if the employer knows of the abnormal susceptibility he may be liable if he deliberately or negligently exposes the employee to that risk, but not if the employee insists on continuing to work and the employer gives the employee full information as to the risk involved.[28–29]

---

[24] In such a case there may well be rights of contribution and indemnity.
[25] See pp. 37–43.
[26] See, *e.g.* Garard v. *A. E. Southey and Co.* [1952] 2 Q.B. 174; *Denham* v. *Midland Employers Mutual Assurance* [1955] 2 Q.B. 437; *Gibb* v. *United Steel Companies* [1957] 1 W.L.R. 668; *O'Reilly* v. *I.C.I.* [1955] 1 W.L.R. 1155, C.A.; *McDermid* v. *Nash Dredging and Reclamation Co. Ltd.* [1987] 3 W.L.R. 212, H.L.
[27] *Smith* v. *Baker* [1891] A.C. 325, H.L.; *Paris* v. *Stepney Borough Council* [1951] A.C. 367; *Davie* v. *New Merton Board Mills* [1959] A.C. 604; *Wilsons and Clyde Coal Co.* v. *English* [1938] A.C. 57, H.L.
[28–29] See *Paris* v. *Stepney Borough Council* [1951] A.C. 367; [1951] 1 All E.R. 42, H.L. (goggles necessary for one-eyed man when not necessary for two-eyed men). *Withers* v. *Perry Chain Co.* [1961] 1 W.L.R. 1314, C.A.; *Kossinski* v. *Chrysler United Kingdom Ltd.* (1974) 15 K.I.R. 225, C.A.; *Cork* v. *Kirby McLean* [1952] 2 All E.R. 402, C.A.

Nature of injury damage, etc., covered by the duty of care

The duty to take reasonable care for the employee's safety means of course that the employer must protect him from liability to injury or the contraction of an industrial disease at work. The employer does not however normally have liability for the safety of the employee's property which he brings on the premises[30–31] though of course the employer may be liable under some other head, *e.g.* if he accepts possession of the employee's goods or deliberately orders an act which may imperil the property of the employee.[32]

Place where duty of care owed

This would normally be the premises of the employer and any outside place of employment. So far as concerns outside places of employment the employer will be liable normally only if he has some measure of control over those outside premises or has been notified of the defects in them and is in a position to ensure that the occupier puts them right.[33]

Duty of employer in respect of negligence of persons other than his servants or agents

This is analogous to the problem discussed above as to the employer's liability for the safety of premises to which he sends his employee but is also affected by the principle that an employer's personal duty of care for the safety of his workmen is said to be a duty which he cannot delegate.[34] In such cases the employer will

---

[30–31] *Deyong* v. *Shenburn* [1946] K.B. 227, C.A.; *Edwards* v. *West Herts Group Hospital Management Committee* [1957] 1 W.L.R. 415, C.A.

[32] Note also that the employer may in certain circumstances be liable for damage, etc., to the employer's property under a specific statutory provision, *e.g.* s.12 of the Offices, Shops and Railways Premises Act 1963—accommodation for clothing—see *McCarthy* v. *Daily Mirror* [1949] 1 All E.R. 801; *C. S. Barr* v. *Cruickshank and Co.* [1959] S.L.T. (Sh. Ct.) 9.

[33] Compare *Smith* v. *Austin Lifts* [1959] 1 W.L.R. 100, H.L.; and *McCloskey* v. *Western Health and Social Services Board* [1983] 4 N.I.J.B., C.A. As to shipbuilding and ship-repairing cases see, *e.g. Mace* v. *R. and H. Green* [1959] 2 Q.B. 14 and *Wingfield* v. *Ellerman's Wilson* [1960] 2 Lloyd's Rep. 16, C.A.

[34] *Cassidy* v. *Minister of Health* [1951] 2 K.B. 343, C.A. and *McDermid* v. *Nash Dredging and Reclamation Co. Ltd.* [1987] 3 W.L.R. 212, H.L.

automatically be liable for injury, etc., caused by the independent contractor, *e.g.* for example a contractor called in to repair dangerous machinery. So far as manufacturers and suppliers of equipment are concerned, they do not constitute independent contractors performing the personal duty of care of the employer[35] but that position has now been altered by the Employer's Liability (Defective Equipment) Act 1969 which enables the employer to be made liable to an employee injured in the course of his employment because of a defect in equipment provided by the employer, even though negligence cannot be established against the employer.[36]

Protection of employee from physical violence by third parties

The employer owes a duty of care to take reasonable precautions to protect his employee from physical violence in the course of the latter performing his duties. This would seem to be so whether the violence is committed by a fellow employee[37] or by strangers.[38] These cases establish that the employer will be liable only if he does not take steps to eliminate a risk which he knows or ought to know is a real risk and not just a mere possibility. An employee who has been subjected to violent treatment on a number of occasions may decide to leave and, in an industrial tribunal, claim "constructive dismissal" against his employer.[39] There is no separate implied term of the employment contract that an employer shall not be *transferred* to dangerous work but only the overall implied term that the employer will provide a safe system of work.[40]

---

[35] *Davie v. New Merton Board Mills* [1959] A.C. 604.
[36] The Act speaks of "equipment" which gives rise to difficulties. It has for example been held in *Coltman v. Bibby Tankers, The Derbyshire* [1987] 3 All E.R. 1068, H.L. that a ship is "equipment." Where the employer is made liable in such circumstances he will normally be able to obtain contribution or indemnity as far as the damages and costs are concerned from the manufacturer of the equipment provided that that manufacturer has not excluded that liability contractually—see Civil Liability (Contribution) Act 1978 and compare *Morris v. Ford Motor Co. Ltd.* [1973] 2 All E.R. 1084, C.A.
[37] *Smith v. Ocean S.S.* [1954] 2 Lloyd's Rep. 482.
[38] See *Charlton v. Forest Printing Ink Co. Ltd.* [1980] I.R.L.R. 331; *West Bromwich Building Society v. Townsend* [1983] I.C.R. 257, *The Times*, January 3, 1983 D.C.; *Keys v. Shoe Fayre Ltd.* [1978] I.R.L.R. 476; *Dutton and Clark v. Daly* [1985] I.C.R. 780.
[39] See Chap. 3.
[40] *Jagdeo v. Smiths Industries* [1982] I.C.R. 47, E.A.T.

(3) STANDARD OF CARE

(a) Generally

The standard of care in general expected of an employer is that of the ordinarily prudent employer.[41] The standard of care is not to be increased by reason of the fact that employers usually insure.[42] Nor is it to be set so high as to become barely distinguishable from the absolute statutory obligations of the employer.[43]

The standard of care required in law falls under three main heads:

 (i) foreseeing the existence of a risk;
 (ii) assessing the magnitude of the risk; and
 (iii) devising reasonable precautions.

As to the first, if the conditions of work contain some element of danger, the employer must foresee the possibility of injury arising therefrom even though it results through the intermediation of an act of inadvertence by the employee and even though that act of inadvertence is of a character which cannot be precisely forecast.[44] For a risk, however, to be reasonably foreseeable, the accident must be one which could or would be reasonably foreseeable "in the ordinary course," as opposed to something which is wholly exceptional or unique, or which no reasonable employer could have been expected to anticipate.[45] As to assessing the magnitude of the risk, there are two factors to be taken into account—the seriousness of the injury risked, and the likelihood of the injury being in fact caused. In addition to this, it may sometimes be necessary to take account of the consequence of not assuming a risk, as, for example, the consequence to the national economy if all trains and vehicles were restricted to a very slow speed. Thus the standard of care owed to a one-eyed man who is employed to hammer rusty bolts will be higher than in the case of a normal man, owing to the greater seriousness of the injury risked if a piece of metal should fly into his eye.[46] In many operations some risk may be unavoid-

---

[41] *Paris* v. *Stepney Borough Council* [1951] A.C. 367; [1951] 1 All E.R. 42, H.L.; *Smith* v. *Austin Lifts* [1959] 1 W.L.R. 100; [1959] 1 All E.R. 81, H.L.; *Vange Scaffolding and Engineering Co.*, *The Times*, March 26, 1987, C.A.
[42] *Davie* v. *New Merton Board Mills* [1959] A.C. 604; [1959] 1 All E.R. 346, H.L.
[43] *Latimer* v. *A.E.C.* [1953] A.C. 643; [1953] 2 All E.R. 449, H.L.
[44] *Thurogood* v. *Van den Berghs and Jurgens* [1951] 2 K.B. 537; *sub nom. Thorogood* v. *Van Den Berghs and Jurgens* [1951] 1 All E.R. 682, C.A.
[45] *Close* v. *Steel Co. of Wales* [1962] A.C. 367; [1961] 3 W.L.R. 319; [1961] 2 All E.R. 953, H.L.
[46] *Paris* v. *Stepney Borough Council* [1951] A.C. 367; [1951] 1 All E.R. 42, H.L.

able, and in such cases as long as the employer takes reasonable care to carry out the operation in a reasonably safe way, his obligation is complete.[47] In one case a factory floor was flooded by an exceptional rain storm; the employers applied sawdust which still left the floor slippery, but the only alternative course was to close the factory altogether; it was held by the House of Lords that the risk was not a sufficiently grave one to justify this exceptional course.[48] It was said in one case that:

> "one must always guard against making industry impossible and against slowing down production by setting unduly high standards or by placing wholly unreasonable requirements on the makers and owners of machinery of this kind."[49]

Nevertheless, although the fact that an operation involves some risk does not necessarily mean that the employer will be liable for any accident which results, the standard of care required is a high one. It is his duty, in considering whether some precautions should be taken against a foreseeable risk, to weigh on the one hand the magnitude of the risk, the likelihood of an accident happening, and the possible seriousness of the consequences if an accident does happen, and on the other hand the difficulty and expense and any other disadvantage of taking the precaution.[50] The risk has always to be balanced against the end, and in this connection there is a considerable difference between a commercial end to make profit and a human end to save life or limb.[51] In considering whether a precaution ought reasonably to have been taken, it will often be important to see what other employers do in respect of similar operations, though this will not necessarily be conclusive.

(b) Abnormal susceptibility of employee

If the employer observes a degree of care which is sufficient to provide reasonable protection for the normal employee, he will not be liable for some injury resulting from the abnormal susceptibility of a particular employee of which he neither knew nor ought to have known. But an employer must be prepared for some higher degree

---

[47] *Newman* v. *Harland & Wolff* [1953] 1 Lloyd's Rep. 114.
[48] *Latimer* v. *A.E.C.* [1953] A.C. 643; [1953] 2 All E.R. 449, H.L.
[49] *Jones* v. *Richards* [1955] 1 W.L.R. 444; [1955] 1 All E.R. 463.
[50] *Morris* v. *West Hartlepool Steam Navigation Co.* [1956] A.C. 552, *per* Lord Reid; *Gawtry* v. *Waltons Wharfingers & Storage Ltd.* [1971] 2 Lloyd's Rep. 489, C.A. Where the possibility of injury is remote, and no better practical method of doing the particular job is known, no liability will arise at common law: *Hindle* v. *Joseph Porritt & Sons Ltd.* [1970] 1 All E.R. 1142.
[51] *Watt* v. *Herts C.C.* [1954] 1 W.L.R. 835; [1954] 2 All E.R. 368, C.A.

of sensitivity than normal, so long as it is not altogether exceptional.[52] If, however, he knew or (probably) ought to have known of the abnormal susceptibility, different considerations apply, and he will ordinarily be under a duty to take extra precautions in relation to that particular employee.[53] An employee who knowingly is suffering from some abnormal susceptibility which exposes him to additional risk of injury is under a duty to disclose this to his employer.[54] If an employee suffers from a known abnormal susceptibility, such as a propensity to dermatitis from grease, and if he wishes despite his knowledge of the risk to continue at a job in greasy conditions, the employer is under no duty to refuse to employ him or to dismiss him from a job involving such conditions, provided that the employer ensures that the employee has a full understanding of the risk involved.[55] It is not clear whether this principle would apply, in addition to those involving a slight risk of minor injury, to those involving a dangerous risk of death, such as, for instance, the employment of a known epileptic on scaffolding.[56]

(c) Guarding against carelessness and indifference on the part of employees

From the reported cases it is not always easy to judge how far an employer is justified in leaving it to his employees to guard against their own carelessness and indifference. He is not entitled to disregard the possibility of slips, errors of judgment, and momentary forgetfulness to which any ordinary employee is subject, but at the same time his obligation does not extend to having to take safety precautions against foolhardiness on the part of the workman. If the work has inherent dangers, such as where machinery is involved, precautions against slips and forgetfulness may well be required which would not otherwise be called for.[57] The same applies where the work is repetitive, rather than varying from job

---

[52] *Board* v. *Thomas Hedley & Co.* [1951] W.N. 422; [1951] 2 All E.R. 431, C.A., *per* Denning L.J.
[53] *Paris* v. *Stepney Borough Council* [1951] A.C. 367; [1951] 1 All E.R. 42, H.L. See also *Porteous* v. *National Coal Board*, 1967 S.L.T. 117; *Cork* v. *Kirby Maclean* [1952] W.N. 399; [1952] 2 All E.R. 402, C.A.; and *Bailey* v. *Rolls Royce (1971) Ltd.* [1984] I.C.R. 688, C.A.
[54] *Cork* v. *Kirby Maclean, supra.*
[55] *Withers* v. *Perry Chain Co.* [1961] 1 W.L.R. 1314; [1961] 3 All E.R. 676, C.A.
[56] *Cork* v. *Kirby MacLean* [1952] 2 All E.R. 402; [1952] 2 T.L.R. 217, C.A.
[57] *Thurogood* v. *Van den Berghs and Jurgens* [1951] 2 K.B. 537; *sub nom. Thorogood* v. *Van Den Berghs and Jurgens* [1951] 1 All E.R. 682, C.A.

to job, for in such cases employees may become careless through their very familiarity with the risk; this is particularly the case in operations involving obvious danger, such as window cleaning or steel erecting.[58]

More care is needed where young people are concerned.[59] On the other hand, where the employee is a skilled and experienced man, the employer will usually be justified in leaving it to him to guard against the possibility of his own carelessness.[60]

An employer is ordinarily entitled to rely on a system whereby employees have to report or remedy defects in their tools which appear in the course of their use.[61]

(d) Giving instructions

For skilled men instructions will not ordinarily be required where the work is within their competence. Instructions for a repeated operation may well be required where none would be called for if it took place on an isolated occasion.[62]

If the danger is obvious to common sense (as, for instance, an untethered and dangerous bull in a loose box), instructions will not be required,[63] nor can an employer be reasonably required to instruct an employee who in fact already knows what to do.[64] But where more than rudimentary knowledge is required to appreciate the dangers, instructions are required for an unskilled labourer, even though they would not be required for a skilled worker who would appreciate the risk from experience.[65]

Where young people are employed in conditions of potential

---

[58] *General Cleaning Contractors* v. *Christmas* [1953] A.C. 180; [1952] 2 All E.R. 1110, H.L.; *Wilson* v. *Tyneside Window Cleaning Co.* [1958] 2 Q.B. 110; [1958] 2 All E.R. 265, C.A.; *Wilson* v. *International Combustion* (July 18, 1956), C.A., noted in [1957] 3 All E.R. 505.
[59] *Woods* v. *Durable Suites* [1953] 1 W.L.R. 857; [1953] 2 All E.R. 391, C.A.; *Wheeler* v. *London and Rochester Trading* [1957] 1 Lloyd's Rep. 69.
[60] *Shanley* v. *West Coast Stevedoring Co.* [1957] 1 Lloyd's Rep. 391, C.A.; *Jenner* v. *Allen West & Co.* [1959] 1 W.L.R. 554; [1959] 2 All E.R. 115, C.A.
[61] *Bastable* v. *Eastern Electricity Board* [1956] 1 Lloyd's Rep. 586, C.A.; *Cf. Smith* v. *Scot Bowyers Ltd.*, *The Times*, April 16, 1986, C.A. (no need for employer to arrange for replacement of employees' worn wellington boots).
[62] *General Cleaning Contractors* v. *Christmas* [1953] A.C. 180; [1952] 2 All E.R. 1110, H.L.
[63] *Rands* v. *McNeil* [1955] 1 Q.B. 253; [1954] 3 All E.R. 593, C.A.
[64] *Jones* v. *A. E. Smith Coggins* [1955] 2 Lloyd's Rep. 17, C.A.
[65] *Payne* v. *Peter Bennie Ltd.* (1973) 14 K.I.R. 395.

danger,[66] or where the work entails some unusual risk,[67] or where the employee is not able to judge the full extent of the risk,[68] some instruction by the employer will usually be required.

If instructions are required, they must be positive,[69] and the duty of giving instructions is not fulfilled merely by telling the employee to read the regulations.[70]

(e) Supervision and enforcement of precautions

The recent trend of decisions has been against imposing too high a duty on the employer in this respect. If, however, the danger is latent rather than patent, as for example the danger of silicosis from foundry dust, constant enforcement may reasonably be necessary in order to make employees wear the masks provided; and enforcement which experienced men would resent may be necessary in the case of young novices.[71] Where a widespread unsafe practice has grown up, such as that of oiling machines in motion, an employer will be liable if, when he knows or ought to know of the practice, he stands by and takes no steps to stop it.[72]

(f) Relevance of previous accidents and complaints

The occurrence or non-occurrence of previous similar accidents is very relevant, and not infrequently conclusive, on the question whether the employer should reasonably have foreseen the likelihood of the accident in question. Even if previous accidents have

---

[66] *Stringer* v. *Automatic Woodturning* [1956] 1 W.L.R. 138; [1956] 1 All E.R. 327, C.A.; *Watts* v. *Empire Transport Co., Ltd.* [1963] 1 Lloyd's Rep. 263.

[67] *Nicolson* v. *Shaw Savill* [1957] 1 Lloyd's Rep. 162. Cf. *White* v. *Holbrook Precision Castings* [1985] I.R.L.R. 215, C.A. (no duty to warn of vibration-induced white-finger).

[68] *Baker* v. *T. E. Hopkins & Son* [1959] 1 W.L.R. 966; sub nom. *Ward* v. *T. E. Hopkins & Sons* [1959] 3 All E.R. 225, C.A., distinguished in *Vinnyey* v. *Star Paper Mills Ltd.* [1965] 1 All E.R. 175. Cf. *Burgess* v. *Thorn Consumer Electronics (Newhaven) Ltd., The Times,* May 16, 1983 (employer's duty to warn employee of danger of tenosynovitis, having had guidance notes from Health and Safety Executive) and *James* v. *Durkin (Civil Engineering Contractors), The Times,* May 25, 1983 (employer's duty to warn driver by notice in cab that lorry was over 12 feet high).

[69] *Baker* v. *Hopkins, supra; Lewis* v. *High Duty Alloy* [1957] 1 W.L.R. 632; [1957] 1 All E.R. 740.

[70] *Barcock* v. *Brighton Corp.* [1949] 1 K.B. 339; [1949] 1 All E.R. 251.

[71] Compare *Woods* v. *Durable Suites* [1953] 1 W.L.R. 857; [1953] 2 All E.R. 391, C.A., with *Clifford* v. *Charles Challen* [1951] 1 K.B. 495; [1951] 1 All E.R. 72, C.A.

[72] *Lewis* v. *High Duty Alloys* [1957] 1 W.L.R. 632; [1957] 1 All E.R. 740.

only resulted in trivial injuries, that may be no reason for failing to foresee that sooner or later such an accident is likely to cause some serious injury.[73]

The absence of previous accidents, although a very material circumstance in rebutting negligence, is by no means conclusive, for it does not follow that one must wait for an accident before a system can be condemned as an unsafe one.[74] Whilst investigation of the accident rate may be relevant in determining whether a system is dangerous, once it is established that a system is dangerous, the fact that there has been no previous accident is of no avail to the employer.[75]

Much the same considerations apply to previous complaints by employees, although their cogency will often be of less weight.

(g) Danger unknown at time of accident

If the danger was one of which the employer neither knew nor ought reasonably to have known, he will not be liable for a resulting accident, for no man can reasonably foresee an accident if the risk is unknown at the time.[76] The degree of knowledge required is that of an ordinarily prudent employer.[77] If a danger is generally known in the industry, or if it has been specifically referred to in bulletins or journals supplied by the trade, knowledge of the danger will be imputed to a defendant employer.[78]

(h) General practice of industry

Since the standard of care required of an employer is that of the ordinarily prudent employer, it is always highly relevant to ask what the general practice of industry was at the time of the accident. If, so far as a fault of omission is concerned, a defendant has observed the general practice in the industry, he will not ordinarily

---

[73] *Kilgollan* v. *W. Cooke & Co.* [1956] 1 W.L.R. 527; [1956] 2 All E.R. 294, C.A.
[74] *Morris* v. *West Hartlepool Steam Navigation Co.* [1956] A.C. 552; [1956] 1 All E.R. 385, H.L.
[75] *Brown* v. *John Mills & Co. (Llanidloes) Ltd.* (1970) 114 S.J. 149, C.A.
[76] *Quinn* v. *Cameron & Robertson*, 1956 S.C. 224; reversed [1958] A.C. 9, H.L.; 1957 S.C., H.L. 22; *Riddick* v. *Weir Housing Corp. Ltd.*, 1971 S.L.T. 24. *Cf. Joseph* v. *Ministry of Defence*, *The Times*, March 4, 1980 ("white finger" not foreseeable from vibrating drills).
[77] *Ibid.*
[78] *Graham* v. *Co-operative Wholesale Society* [1957] 1 W.L.R. 511; [1957] 1 All E.R. 654. *Cf. Bryce* v. *Swan Hunter Group*, *The Times*, February 19, 1987 (employers' knowledge of risks of asbestos).

be guilty of negligence, unless it is shown that the omission was something which was clearly unreasonable or imprudent.[79] Similarly, if a plaintiff can show that an omission by his employer was contrary to the general practice of the industry, the employer will have to provide some clear and strong reason in order to escape liability.

(i) Relevance of statutory safety provisions to liability at common law

If the circumstances of an accident are just outside the scope of the statutory provisions, the provisions will be irrelevant to the question of liability at common law.[80] In other respects, however, the existence or absence of a particular safety provision in a code for a particular trade may be of some relevance to the standard of care required at common law. Thus knowledge of the danger of pneumoconiosis in certain grinding operations has been presumed from the existence of regulations devised to protect certain classes of workmen; the plaintiff itself was not within these classes, but nevertheless succeeded at common law.[81] Similarly, the existence of mining regulations against premature shot firing will assist to establish liability at common law in respect of such conduct in a case where the regulations for technical reasons do not apply.[82]

Conversely, an employer who has discharged to the full his statutory obligations may be able to claim that he has also thereby fulfilled the standard of care required of him at common law.[83] Nevertheless, the employer may fail to provide a safe system of work notwithstanding that his statutory duty in the particular case has been complied with. Thus it has been held not to be a breach of the Factories Act, 1961, for an experienced man to lift 145 lbs. Yet the employer was liable in negligence, for his "system" of

---

[79] *Cf. Brown* v. *John Mills & Co. (Llanidloes) Ltd.* (1970) 114 S.J. 149, C.A., where it was said that "no one could claim to be excused for want of care because others were as careless as himself." In *Grioli* v. *Allied Building Co. Ltd., The Times*, April 10, 1985, C.A., it was held that as "there was no significant practice" for carpenters to use gauntlets to protect against cuts by glass, the employer was not liable for failure to provide them.
[80] *Chipchase* v. *British Titan Products* [1956] 1 Q.B. 545; [1956] 1 All E.R. 613, C.A.
[81] *Franklin* v. *Gramophone Company* [1948] 1 K.B. 542; [1948] 1 All E.R. 353.
[82] *N.C.B.* v. *England* [1954] A.C. 403; [1954] 1 All E.R. 546, H.L.; *Cf. Hartley* v. *Mayoh* [1954] 1 Q.B. 383; [1954] 1 All E.R. 375, C.A.
[83] *Caulfield* v. *Pickup* [1941] 2 All E.R. 510; *Franklin* v. *Gramophone Co., supra*; *Qualcast* v. *Haynes* [1959] A.C. 743; [1959] 2 All E.R. 38.

leaving the individual employee to decide whether he needed assistance did not discharge the duty of care to the plaintiff.[84]

(j) Ordinary risks of service

Many trades and industries require the performance of operations which carry an unavoidable and inherent risk, sometimes called an ordinary risk of the service, against which the employer cannot reasonably be expected to protect his employees. But if an operation entails a real, serious and unnecessary risk, an employer may not be able to excuse himself on the ground that it is one which is ordinarily found or accepted in the trade. Thus it is negligent for a master window cleaner to accept the view that the job of a window cleaner is inherently risky, and to give no further consideration to methods of making it safer.[85]

(k) Work carried on on premises of third party

Where an employer sends his employees to work on premises occupied by a third party, he still owes his employees a duty to use reasonable care, but what is reasonable will vary with the circumstances, and one importance circumstance in this respect is whether the premises are under the control of the employer.[86-87] Thus, where the handle of a window in a private house came away in the hands of a window cleaner who had frequently cleaned the windows before, it was held that the employer was not liable for not having taken steps to detect or obviate the danger.[88] In another case,[89] a plaintiff lift repairer was injured when a defective door gave way on the premises of a third party. He had several times reported to his employers that the door was defective, but on the occasion of the accident the defect which developed was of a different type from the one which he had reported.

---

[84] *Kinsella* v. *Harris Lebus Ltd.* (1964) 108 S.J. 14, C.A.
[85] *Drummond* v. *British Building Cleaners* [1954] 1 W.L.R. 1434; [1954] 3 All E.R. 507, C.A.
[86-87] *Wilson* v. *Tyneside Window Cleaning* [1958] 2 Q.B. 110; [1958] 2 All E.R. 265, C.A.
[88] *Wilson* v. *Tyneside Window Cleaning, supra*, expressly approved by the House of Lords in *Smith* v. *Austin Lifts* [1959] 1 W.L.R. 100; [1959] 1 All E.R. 81, H.L.
[89] *Smith* v. *Austin Lifts, supra*.

(4) PROOF OF NEGLIGENCE

(a) Generally

Ordinarily the onus of proof in any claim lies on the plaintiff.[90] In an action for negligence, the standard of proof required is that a plaintiff should, on the evidence, establish a balance of probability in his favour.[91] If he does this, he discharges the burden of proof, but if the evidence viewed as a whole leaves the balance of probabilities more or less equal either way, his claim will fail.

In an ordinary claim the matters to be proved fall under three heads: first, the way the accident happened, secondly, that what happened constituted negligence, and, thirdly, that injury or damages was caused as a result of the negligence.

(b) Proving how the accident happened

If the evidence fails to explain or leaves in doubt how the accident happened, the plaintiff will find it difficult to establish liability. In one case[92] a lineman going up an electric pole was electrocuted on reaching the tie bar. The crucial question was as to how his body had come into contact with the current, and as the evidence failed to explain this, the claim failed. But, even though direct evidence of how an accident happened may be lacking, it may still be possible to infer this as a matter of probability on the evidence as a whole.[93] If a plaintiff proves that an accident happened by reason of some defect which would ordinarily not be present, he ordinarily will not be required to take the chain of proof back further in order to explain how the defect arose, provided that there is some

---

[90] As to what constitutes an admission of liability by the defendant, absolving the plaintiff from the need to prove negligence, or damage resulting therefrom, see *Rankine* v. *Garton Sons & Co. Ltd.* [1979] 2 All E.R. 1185, C.A. To prove his case, the plaintiff may require information from the defendant, *e.g.* by order for discovery of documents—see, for example, *Drennan* v. *Brooke Marine Ltd., The Times*, June 21, 1983 and *Kirkup* v. *British Rail Engineering Ltd.*, C.A. [1983] 1 W.L.R. 1165; [1983] 3 All E.R. 147, C.A.; *Aston* v. *Firth Brown* [1984] 5 C.L. 142 (discovery, etc., in occupational deafness cases) and *Thorne* v. *Strathclyde Regional Council* (O.H.), 1984 S.L.T. 161 (order for inspection of machine).
[91] *Miller* v. *Ministry of Pensions* [1947] W.N. 241; [1947] 2 All E.R. 372.
[92] *Youngman* v. *Pirelli* [1940] 1 K.B. 1; 109 L.J.K.B. 420, C.A.
[93] *Clowes* v. *N.C.B., The Times*, April 23, 1987, C.A. (Court took account of habitual careless behaviour and presumed accident due to it).

evidence, direct or inferential, to show that the defect was within the responsibility of his employers. In such cases, it will be for the defendants to show, if they can, that the defect arose in circumstances not amounting to negligence.

(c) Res ipsa loquitur (The matter speaks for itself)

Closely connected with the above is the principle embodied in the phrase *res ipsa loquitur*. Under this principle, if a plaintiff shows that he has suffered injury or other damage which has been caused by some event or thing:

(a) which at all material times was under the sole control of the defendants, and
(b) which would not ordinarily occur or exist without negligence on the part of the person in control, then he is entitled to succeed, unless the defendant can affirmatively show that there was no negligence.

Where a plaintiff was employed to immerse objects in a gas-heated tank which suddenly blew up, it was held that the plaintiff was entitled to succeed on the principle *res ipsa loquitur*, although there was no evidence as to how the explosion had occurred.[94] A brick falling from a building under repair has been held to be *res ipsa loquitur*, though in this case the defendants were able to establish that there had been no negligence on their part.[95] Where a scaffolding plank gave way for no explained reason and the plaintiff fell, he was held to be able to recover damages from the building owner.[96]

Where an event or thing which would constitute *res ipsa loquitur* if under the sole control of a defendant is under the joint control of two or more defendants, there is some authority for saying that each defendant must show that there was no negligence on his part in order to escape liability,[97] but this is not altogether certain.

---

[94] *Moore* v. *Fox* [1956] 1 Q.B. 596, C.A.; *Bennett* v. *Chemical Construction (G.B.) Ltd.* [1971] 1 W.L.R. 1571; [1971] 3 All E.R. 822.
[95] *Walsh* v. *Holst* [1958] 1 W.L.R. 800; [1958] 3 All E.R. 33, C.A.
[96] *Kealey* v. *Heard* [1983] 1 All E.R. 973.
[97] *Roe* v. *Ministry of Health* [1954] 2 Q.B. 66; [1954] 2 All E.R. 131, C.A.; *Baker* v. *Market Harborough Industrial Co-operative Society* [1953] 1 W.L.R. 1472; 97 S.J. 861, C.A.

Res ipsa loquitur cannot apply where the cause of the accident is known.[98] Where *res ipsa loquitur* does apply, the defendant must offer an explanation of how the accident probably occurred, and further show that such an explanation is consistent with absence of negligence on his part.[99]

(d) Proof of causation

In order to succeed in a claim for damages, a plaintiff must not only prove that the defendant was negligent, but that the negligence was the cause of the injuries complained of. Where there are two or more causes of an accident, it is clearly settled that a wrongdoer cannot excuse himself by pointing to another cause. It is enough that the wrong done by the wrongdoer should be a cause, and it is unnecessary to evaluate competing causes and ascertain which of them is dominant.[1] If, however, the other cause is the contributory negligence of the plaintiff himself, the damages awarded to a plaintiff will be reduced by a percentage appropriate to the degree of his own negligence.[2]

Where an employee contracts a disease which could have arisen from independent causes or which alternatively could have been caused by exposure to conditions of work for which the employer would be liable if there were proof that damage resulted therefrom, there is a prima facie presumption that the disease was caused by those conditions.[3]

In a case where a plaintiff sustains unusual or unexpected injuries or damage, the question may arise whether such injuries or damage can fairly be said to have been caused by the accident.

---

[98] *Barkway* v. *South Wales Transport* [1950] A.C. 185; [1950] 1 All E.R. 391, H.L.; *Bolton* v. *Stone* [1951] A.C. 850; [1951] 1 All E.R. 1078, H.L.
[99] *Woods* v. *Duncan* [1946] A.C. 401; [1946] 1 All E.R. 420, H.L.; *Turner* v. *N.C.B.* (1949) 65 T.L.R. 580, H.L.; *Walsh* v. *Holst* [1958] 1 W.L.R. 800; [1958] 3 All E.R. 33, C.A.; *Colvilles Ltd.* v. *Devine* [1969] 1 W.L.R. 475; [1969] 2 All E.R. 53; 1969 S.C., H.L. 69.
[1] *Heskell* v. *Continental Express* [1950] W.N. 210.
[2] See p. 34.
[3] *Gardiner* v. *Motherwell Machinery & Scrap Co.* [1961] 1 W.L.R. 1424; [1961] 3 All E.R. 831, H.L.; 1962 S.L.T. 2; 1961 S.C., H.L. 1 *cf. Hotson* v. *East Berkshire Area Health Authority* [1987] 3 W.L.R. 232, H.L. (no damages for loss of 25 per cent. chance of avoiding disease, as defendant's negligence not cause of disease). *Cf. Wilsher* v. *Essex Area Health Authority, The Times*, March 11, 1988, H.L.

Such damage is only recoverable if it is of such a kind as the reasonable man should have foreseen.[4] Where an employee was struck on the lip by a piece of molten metal, so that cancer ensued at the site of the burn from which the employee died, it was held that the employers were liable, as the test was not whether they could have foreseen that the burn would cause cancer, but whether they could have foreseen the type of danger.[5]

Where, subsequently to the accident, some further event takes place which aggravates the damage, the rule is that if the new event constitutes a new cause, the damage resulting therefrom will not be recoverable. To constitute a new cause, the event must be a new cause coming in disturbing the sequence of events, something that can be described as either unreasonable or extraneous or extrinsic.[6] It is sometimes said that, to constitute a new cause in this sense, the event must be one which breaks the chain of causation. Negligent or inefficient medical treatment may constitute such a new cause.[7] But a case where ships collided at sea, and the captain of one ship ordered the men to the boats, and owing to the rough sea one of the boats capsized, it was held that there was no break in the chain of causation from the collision.[8] The test in such cases is whether the plaintiff's subsequent conduct was or was not reasonable.[9]

---

[4] *Overseas Tankship (U.K.)* v. *Morts Dock & Engineering Co. (The Wagon Mound)* [1961] A.C. 388, P.C. See also *Doughty* v. *Turner Manufacturing Co., Ltd.* [1964] 1 Q.B. 518, C.A. There an asbestos cement lid was negligently dropped into a cauldron of molten sodium cyanide. But the resulting explosion was not reasonably foreseeable. The Court of Appeal held that the employers were not liable. Contrast *Bradford* v. *Robinson Rentals Ltd.* [1967] 1 W.L.R. 337; liability for frostbite resulting when employee obliged to drive a long distance in an unheated vehicle, even though the cold weather was unusual in this country: the kind of injury was foreseeable even though its precise nature and extent were not; *Stewart* v. *West African Terminals Ltd.* [1964] 2 Lloyd's Rep. 371, where the involuntary act of the plaintiff in reaction to the dangerous situation created by his employers' negligence did not make his eventual injury too remote a consequence of their negligence. Hence they were held to be liable.
[5] *Smith* v. *Leech Brain & Co.* [1962] 2 Q.B. 405. See also *Curran* v. *Scottish Gas Board*, 1970 S.L.T. (Notes) 33; *Robinson* v. *Post Office* [1974] 2 All E.R. 737, C.A. Cf. *Pigney* v. *Pointer's Transport* [1957] 1 W.L.R. 1121; [1957] 2 All E.R. 807 (suicide); *Tremain* v. *Pike* [1969] 1 W.L.R. 1556.
[6] *The Oropesa* [1943] 1 All E.R. 211, C.A., *per* Lord Wright.
[7] *Rothwell* v. *Caverswall Stone Co.* [1944] 2 All E.R. 350, C.A.; *Hogan* v. *Bentinck Collieries* [1949] W.N. 109; [1949] 1 All E.R. 588, H.L.
[8] *The Oropesa, supra.*
[9] *A. C. Billings* v. *Riden* [1958] A.C. 240; [1957] 3 All E.R. 1, H.L. See also *McKew* v. *Holland & Hannen & Cubitts (Scotland) Ltd.*, 1970 S.C., H.L. 20; [1969] 3 All E.R. 1621, H.L.

(5) SYSTEM OF WORK

(a) Generally

One important aspect of the employer's duty of care towards his employees is his duty to take reasonable care to lay down a safe system of work. In former days, when common employment was a defence to casual acts of negligence by a fellow employee, but not to a fault in the system of work, the dividing line between faults in system and other acts of negligence was of high importance, but today this technical distinction is of no significance, and the only relevance of system of work as a phrase is that it conveniently sums up one important aspect of the general duty of an employer to use reasonable care.

One question in this connection is as to when an employer is reasonably required to lay down a system of work. Thus in one case an accident occurred while the load of a lorry was being lashed down, and the question arose whether this was a case where some system of work should have been imposed in the interests of safety. It was held that it was not, and the claim failed.[10] In this case Lord Reid said:

"A system of working normally implies that the work consists of a series of similar or somewhat similar operations and the conception of a system of working is not easily applied to a case where only a single act of a particular kind is to be performed. Recently, however, this obligation has been extended to cover certain cases where only a single operation is involved. I think the justification for this is that, where an operation is of a complicated or unusual character, an employer careful of the safety of his men would organise it before it was begun and in that sense provide a safe system of working for it. Where such organisation is called for, the employer must provide it, and he cannot delegate his responsibility for it, but cases in which such a duty has been found to exist are comparatively few and it has never been suggested that such an obligation arises in every case where a group of the employer's servants are doing some work which may involve danger if negligently performed."

In another case, an employee employed by one repairer of a number repairing a ship fell through an open tank top. Although

---

[10] *Winter* v. *Cardiff R.D.C.* [1950] W.N. 193, H.L. *Cf. Hayes* v. *N.E. British Road Services* [1977] 7 C.L. 173 (employer's duty to promote safe system of unloading lorry).

the defendant employers were not the occupiers nor in control of the ship, they were held liable, because "in this there was no system or method at all. What was everybody's business became nobody's business. . . . No one exercised any co-ordinating function or final supervision to see that the place was safe to work in."[11]

If an employee is sent to work on premises which are not under the control of the employer, the latter must exercise all the more care in regard to his system of working.[12] Furthermore, although in an ordinary case an employer may be justified in leaving the system of work to his employees if they are men of skill and experience, this may not be the case where the operation is one of "proved danger," such as window cleaning and climbing ladders.[13] Where an employer instructs his workmen in a method of work that is dangerous, he is failing in his duty to provide a safe system of work.[14]

(b) In particular cases

The dangerous operation of window cleaning has been the subject of a number of decisions, which serve to illustrate how far an employer is under a duty to lay down a system of work for operations which involve obvious danger. In one case, a window cleaner fell when a sash which provided the only available handhold suddenly closed, and it was held by the House of Lords, on the evidence adduced, that the employers were not negligent in failing to provide ladders or hooks for safety belts, but that they were negligent in failing to have a system whereby window cleaners employed by them were instructed to test windows and always to see that one sash was kept open.[15] In a subsequent case, the employers were held liable for failing to have a system whereby their employees were instructed in what was known as the transom method, *i.e.* attaching a safety belt to the transom.[16] But, in a case where a defective sash handle suddenly came away,

---

[11] *Donovan* v. *Cammell Laird* [1949] 2 All E.R. 82, *per* Devlin J.
[12] *Wilson* v. *Tyneside Window Cleaning* [1958] 2 Q.B. 110, C.A., *per* Parker L.J., approved by Lord Denning in *Smith* v. *Austin Lifts* [1959] 1 W.L.R. 100, H.L.
[13] *Prince* v. *Carrier Engineering Co.* [1955] 1 Lloyd's Rep. 401.
[14] *Herton* v. *Blaw Knox Ltd.* (1968) 6 K.I.R. 35.
[15] *General Cleaning Contractors* v. *Christmas* [1953] A.C. 180; [1952] 2 All E.R. 1110, H.L.
[16] *Drummond* v. *British Building Cleaners* [1954] 1 W.L.R. 1434; [1954] 3 All E.R. 507, C.A.

it was held that failure by the employers to lay down a system in this respect did not make them liable.[17] From these decisions it may be seen that it may not always be easy to deduce when, in connection with operations of obvious danger, the employer will in law be required to lay down a system of work for skilled and experienced employees.

A class of injury which not infrequently arises is that sustained by a man when lifting or moving a heavy object. In two unreported decisions of the Court of Appeal it has been held to be an unsafe system if an employee is required to lift a weight of 420 pounds unaided, or to manhandle an object weighing 500 pounds. On the other hand the plaintiff has failed in the following cases: lifting bundles of canes weighing 84 pounds,[18] lifting a ladder weighing 65 pounds,[19] lifting a weight of 260 pounds with two men,[20] moving three-hundredweight blocks of pig iron by means of hooks and crowbars but without a winch,[21] and (in unreported decisions of the Court of Appeal) lifting an object of 200 pounds with two men, and similarly lifting an object of one-and-a-half hundredweight. It has been said that where a workman is required to lift articles of different sizes and weights, he should not be left to decide when he needs assistance.[22] In the circumstances of the case, though, it was held to be safe to employ a one-legged man on a lifting job and tell him to ask for assistance when he needed it, since he was only lifting chains weighing 65 pounds, and assistance was always close at hand when required. Failure to ensure co-ordination of the activities of several workmen engaged in a lifting operation constitutes an unsafe system of work.[23]

As in all cases where the standard of care required of an employer is in question, that standard will be that of the ordinarily prudent employer, and this applies no less to questions involving system of work than to other cases. While it is for the employer and not the employee to devise and provide a safe system, in a claim made on the basis of failure to provide a safe system of work, it may be necessary for the plaintiff to show some practicable

---

[17] *Wilson* v. *Tyneside Cleaning Co.* [1958] 2 Q.B. 110; [1958] 2 All E.R. 265. C.A.
[18] *Pipe* v. *Chambers Wharf* [1952] 1 Lloyd's Rep. 194.
[19] *Hegarty* v. *Rye-Arc* [1953] 1 Lloyd's Rep. 465.
[20] *Pasfield* v. *Hay's Wharf* [1954] 1 Lloyd's Rep. 150.
[21] *Gray* v. *Vickers-Armstrongs (Shipbuilders), Ltd.* [1963] 1 Lloyd's Rep. 143.
[22] *Per* Widgery L.J. in *Peat* v. *N.J. Muschamp & Co. Ltd.* (1970) 7 K.I.R. 469 at 476–477, C.A. Cf. *Black* v. *Carricks (Caterers) Ltd.* [1980] I.R.L.R. 448, C.A. (not negligent to leave shop manageress to ask for customers' help, when assistant away ill).
[23] *Upson* v. *Temple Engineering (Southend)* [1975] K.I.L.R. 171.

alternative to the system of work which was operative at the time of the accident.[24]

(6) PLACE OF WORK (INCLUDING ACCESS)

(a) Generally

One important aspect of the duty of an employer to take reasonable care for the safety of his employees is his duty to take reasonable care to provide a safe place of work, which in this context includes safe means of access.[25-26] In the ordinary case the place of work and the means of access to it will be under the control of the employer, but where this is not so different considerations may apply. What is a reasonably safe place of work is a question of fact in each case which peculiarly depends on the circumstances.[27] It has been said that where an employer expects his workmen to employ a very old-fashioned method to do a particular job, it is more important than ever that the duty to provide a safe place of work should be complied with.[28]

The duty of the employer with regard to a safe place of work includes means of access to and from the workplace and stairs and approaches generally,[29] and it may well also include a duty to supervise the means by which the general body of workpeople enter or leave their place of work. For instance, if an employee is injured through being pushed by an uncontrolled surge of other employees leaving a workroom, the employers may well be liable.[30] If an employer provides two means of access each of which is reasonably safe, he is not bound to tell his employees to use the safer of the two.[31]

---

[24] See *Colfar* v. *Coggins & Griffith (Liverpool) Ltd.* [1945] A.C. 197, H.L.; *Dixon* v. *Cementation Ltd.* [1960] 3 All E.R. 417, C.A.; *General Cleaning Contractors Ltd.* v. *Christmas* [1953] A.C. 180, H.L.; and *Gilfillan* v. *N.C.B.*, 1972 S.L.T.(Sh.Ct.) 39.

[25-26] *Hurley* v. *Sanders* [1955] 1 W.L.R. 470; [1955] 1 All E.R. 833.

[27] See *Parker* v. *Vickers Ltd.* [1979] 10 C.L. 161 (wrong to expect same high standard in ship undergoing sea trials as in a factory).

[28] *Busby* v. *Robert Watson (Constructional Engineers) Ltd.* (1973) 13 K.I.R. 498.

[29] *Light* v. *Bourne & Hollingsworth* [1963] C.L.R. 2412; *Stewart* v. *West African Terminals, Ltd.* [1964] 1 Lloyd's Rep. 409; *Hasley* v. *South Bedford Council, The Times*, October 18, 1983.

[30] *Lee* v. *John Dickinson*, (1960) 110 L.J. 317; [1960] 5 C.L. 370; *Bell* v. *Blackwood Morton & Sons*, 1960 S.C. 11; 1960 S.L.T. 145. See also *Lazarus* v. *Firestone Tyre & Rubber Co.* [1963] C.L.Y. 2372.

[31] *Fowler* v. *British Railways Board* [1969] 1 Lloyd's Rep. 231, C.A.

(b) Places of work

Slipping and tripping on the approaches to and the floors of factory premises appears to constitute a hazard which is frequently litigated. Thus there has been held to be no liability in respect of a loose plank on a building site,[32] or of a floor slippery from grease and rainwater,[33] or of an icy surface resulting from a sudden snowfall,[34] or of débris on the floor of a factory.[35] On the other hand, the employers were held liable where there was a slippery duckboard at the site of a water tap,[36] where there was a patch of oil on a loading bank adjoining an office where the plaintiff worked,[37] and where there was a patch of slippery ice in a cold store.[38]

Work carried on without staging at a height of eight feet has been held to be safe at common law,[39] but this may not be so in a case where the Building Regulations apply. Staging six feet high and nine inches wide has also been held to be safe.[40] On the other hand, roof work, where the plaintiff had to kneel nine inches from an unguarded edge, has been held unsafe.[41] With regard to work at a height generally, it has been held to be unsafe if a scaffold has no guard rail,[42] if staging is liable to swing,[43] if the only access to steel framework is along girders,[44] and where a scaffold had a "trap end."[45]

Some more unusual cases where liability has been established include: combustible material lying near a boiler,[46] a hole dug in the middle of a farmyard,[47] an artificial cold draught in a factory,[48] a ceiling with beams only four feet eight inches high,[49] a scaffold

---

[32] *Field* v. *Perrys* [1950] W.N. 320; [1950] 2 All E.R. 521.
[33] *Davies* v. *De Havilland* [1951] 1 K.B. 50; [1950] 2 All E.R. 582.
[34] *Thomas* v. *Bristol Aeroplane Co.* [1954] 1 W.L.R. 694; [1954] 2 All E.R. 1, C.A. Cf. Factories Act 1961, s.29 and cases thereon, *e.g.* Woodward v. Renold Ltd. [1980] I.C.R. 387 and *Darby* v. *G.K.N.* [1986] I.C.R. 1.
[35] *Prince* v. *Carrier Engineering* [1955] 1 Lloyd's Rep. 401; *Stanley* v. *Concentric (Pressed Products) Ltd.* (1972) 11 K.I.R. 260, C.A.
[36] *Davidson* v. *Handley Page* [1945] 1 All E.R. 235; 61 T.L.R. 178, C.A.
[37] *Mahoney* v. *Hay's Wharf* [1963] 2 Lloyd's Rep. 312.
[38] *McDonald* v. *B.T.C.* [1955] 1 W.L.R. 1323; [1955] 3 All E.R. 789.
[39] *Reading* v. *Harland & Wolff* [1954] 1 Lloyd's Rep. 131, C.A.
[40] *Chipchase* v. *British Titan* [1956] 1 Q.B. 545; [1956] 1 All E.R. 613, C.A.
[41] *Harris* v. *Brights Contractors* [1953] 1 Q.B. 617; [1953] 1 All E.R. 395.
[42] *Pratt* v. *Richards* [1951] 2 K.B. 208; [1951] 1 All E.R. 90n.
[43] *Taylor* v. *Ellerman Wilson* [1952] 1 Lloyd's Rep. 144.
[44] *Sheppey* v. *Matthew T. Shaw & Co.* [1952] W.N. 249; [1952] 1 T.L.R. 1272.
[45] *Simmons* v. *Bovis* [1956] 1 W.L.R. 381; [1956] 1 All E.R. 736.
[46] *D'Urso* v. *Sanson* [1939] 4 All E.R. 26.
[47] *Rowden* v. *Gosling* [1953] C.P.L. 218; [1953] C.L.Y. 2420.
[48] *Murray* v. *Walnut Cabinet Works*, 105 L.J. 41; [1954] C.L.Y. 1320, C.A.
[49] *Gemmell* v. *P.L.A.* [1955] 1 Lloyd's Rep. 5, C.A.

with a "trap end,"[50] an inadequate fire exit,[51] and work on an asbestos roof without crawling boards.[52]

(7) PLANT AND APPLIANCES

(a) Generally

Another aspect of the duty of an employer is his duty to use reasonable care to provide safe plant and appliances. Where safe plant is available, the employer will not be in breach of duty to an employee whose authority includes the duty to select the plant to be used.[53] So far as machinery is concerned, his duty in the vast majority of cases will be found in the various statutory provisions, but in cases where this is not so, a similar duty may well exist at common law.

Where an employer designs a piece of apparatus, he owes a duty to his employees to design it competently. If when put to work it works in an unpredictable and dangerous fashion, he owes a further duty to tell the employees working on it that they must take special precautions, and to indicate what are the proper ones to take.[54]

(b) In particular cases

In two reported cases the employer has been held negligent in leaving his employee to fend for himself in trying to find or borrow suitable equipment.[55] An employer will not be liable for latent defects by which is meant a defect which is not discoverable by the exercise of reasonable care, as, for instance, when the connecting rod of a machine suddenly breaks.[56] An employer will be liable for an injury caused by a defect which was patent and would have been evident on inspection.[57]

---

[50] *Simmons* v. *Bovis* [1956] 1 W.L.R. 381; [1956] 1 All E.R. 736.
[51] *Nicholls* v. *Reemer*, 107 L.J. 378; [1957] C.L.Y. 2351.
[52] *Jenner* v. *Allen West & Co.* [1959] 1 W.L.R. 554; [1959] 2 All E.R. 115, C.A.
[53] *Richardson* v. *Stephenson Clarke Ltd.* [1969] 1 W.L.R. 1695; [1969] 3 All E.R. 705.
[54] *McPhee* v. *General Motors Ltd.* (1970) 8 K.I.R. 885.
[55] *Lovell* v. *Blundells* [1944] K.B. 502; [1944] 1 All E.R. 53; *Graves* v. *J. & E. Hall* [1958] 2 Lloyd's Rep. 100.
[56] *Roberts* v. *T. Wallis* [1958] 1 Lloyd's Rep. 29.
[57] *Baxter* v. *St. Helena Group Hospital Management Committee*, *The Times*, February 15, 1972.

With regard to defective equipment, the law was changed by the Employer's Liability (Defective Equipment) Act 1969.[58] In *Davie* v. *New Merton Board Mills*,[59] the House of Lords established that if a tool was inherently defective, the employer would not be liable if it was a standard tool which he had bought through reputable sources, for in obtaining the tool in that way, he would have done all that could reasonably be required of him. Under the Act, an employee who suffers personal injury in the course of his employment because of a defect in equipment[60] provided by his employer for the purposes of his business, can recover damages from his employer even though no negligence can be proved against the employer as long as the defect is attributable in whole or in part to the fault of a third party, who need not be identified.[61]

(8) PROTECTIVE EQUIPMENT

(a) When protective equipment must be provided

If an employee is suffering from some known disability which makes him abnormally susceptible to either a greater risk of injury or to a risk of greater injury, added care in the provision of protective equipment will be called for. Thus, where an employee is employed to hammer rusty bolts under the chassis of a vehicle, the risk ordinarily of a piece of metal getting into the eye may not be sufficiently great to warrant the provision of goggles, but if the employee is a one-eyed man, an employer failing to provide goggles will probably be liable in the event of accident.[62] The question whether goggles are required for a particular operation is one that has frequently arisen, and the criterion in each case is whether or not there is a likelihood, as opposed to a mere exceptional possibility, that substantial damage to an eye may sooner or later arise. Thus, it has been held not to be negligent not to issue goggles for the work of breaking concrete with a mechanical

---

[58] c. 37.
[59] [1959] A.C. 604; [1959] 21 W.L.R. 331; [1959] 1 All E.R. 346, H.L. *Cf. Marston* v. *British Railways Board* [1976] I.C.R. 353, C.A.
[60] "Equipment" includes the entirety of a ship—*Coltman* v. *Bibby Tankers Ltd.* [1987] 3 All E.R. 1068, H.L.
[61] Employer's Liability (Defective Equipment) Act 1969, s.1(1).
[62] *Paris* v. *Stepney Borough Council* [1951] A.C. 367; [1951] 1 All E.R. 42, H.L.

pick,[63] or for sweeping the wall of a dry dock,[64] and similarly (in unreported decisions of the Court of Appeal) for the following operations: planing railway wheels, removing alum, holding a large spanner while it is struck by a fellow employee, removing metal burrs by mechanical means, carrying lime putty, and breaking coal. On the other hand, goggles have been required for cutting steel piping,[65] chipping a brick wall,[66] operating a carborundum wheel,[67] and (in unreported decisions of the Court of Appeal) for chiselling brass inserts out from metal, and for the pneumatic drilling of concrete. Even if goggles are not required, an employer may nevertheless be negligent if he does not provide an employee who always wears spectacles with special safety-spectacles.[68] The need to provide barrier cream as a protection against material which are likely to cause industrial dermatitis has also been the subject of a number of decisions. Thus, in two cases it has been held that barrier cream is required for the manufacture of synthetic glue,[69] but recently the value of barrier cream, at any rate in operations involving cement, oil and grease, has been doubted, as being presumptive only, and not yet proved scientifically.[70]

Where employees carry molten metal, the employer should have safety spats available for those who want to wear them.[71]

It is now established that an employer who fails to provide adequate protection against noise, in the form of ear plugs or muffs, and, in appropriate cases, soundproofing of walls etc., will be in breach of his common law duty of care to his employees. This was established in a case where exposure to excessive noise caused

---

[63] *Walsh* v. *Allweather Mechanical Grouting Co.* [1959] 2 Q.B. 300; [1959] 2 All E.R. 588.
[64] *Johnson* v. *Cammell Laird & Co., Ltd.* [1963] 1 Lloyd's Rep. 237.
[65] *Paling* v. *A. Marshall (Plumbers)* (November 22, 1957), unreported; [1957] C.L.Y. 2420.
[66] *Welsford* v. *Lawford Asphalte Co.* (May 16, 1956), unreported; [1956] C.L.Y. 5984.
[67] *Nolan* v. *Dental Manufacturing Co.* [1958] 1 W.L.R. 936; [1958] 2 All E.R. 449.
[68] *Pentney* v. *Anglian Water Authority* [1983] I.C.R. 464 (employer negligent, even though not a job to which the Protection of Eyes Regulations 1974 applied).
[69] *Clifford* v. *Charles Challen* [1951] 1 K.B. 495; [1951] 1 All E.R. 72, C.A.; *Woods* v. *Durable Suites* [1953] 1 W.L.R. 857; [1953] 2 All E.R. 391, C.A.
[70] *Watson* v. *Ready Mixed Concrete, The Times,* January 18, 1961. (*Cf. Brown* v. *Rolls Royce* [1960] 1 W.L.R. 210; [1960] 1 All E.R. 577; 1960 S.C., H.L. 22.) or a memorandum on the prevention of industrial dermatitis from synthetic resins, see F 331 (H.M.S.O.).
[71] *Qualcast (Wolverhampton)* v. *Haynes* [1959] A.C. 743; [1959] 2 All E.R. 38, H.L.

the employee to go deaf.[72] But the view that no duty can exist unless noise produces a risk of deafness has been rejected.[73] In certain stringently defined circumstances, occupational deafness is a prescribed disease for the purposes of Industrial Injuries benefits under the social security legislation.[74] A lesser degree of deafness than that prescribed by the regulations may nevertheless ground a common-law action for damages.[75]

In general, the test of whether a particular item of protective equipment should be provided is merely one aspect of the general duty of an employer to use reasonable care, the standard of such duty being set by the standard of care which an ordinarily prudent employer would take, having regard to the foreseeability of the risk, the magnitude of the risk itself, and the practicability of taking effective precautions. In so far as what constituted reasonable care must necessarily vary from case to case, the above illustrations must be regarded as illustrations only, rather than as constituting any wholly reliable guide to the likely outcome of a similar issue.

(b) What constitutes provision

Where an operation calls for some item of protective equipment, it is the duty of the employer to provide it.[76] What constitutes a providing may vary from case to case, but, in the case of goggles, they should either be put in a place where they come easily and obviously to hand, or alternatively the employee should be given

---

[72] *Berry* v. *Stone Manganese & Marine Ltd.* [1972] 1 Lloyd's Rep. 182; for a comprehensive review of the law, and types of safety precautions, see *McCafferty* v. *Metropolitan Police District Receiver* [1977] 2 All E.R. 766, C.A. (police officer forced to retire prematurely due to damage to hearing by testing guns—employers liable—no contributory negligence by officer in not asking for, *e.g.* soundproofing) and *Thompson* v. *Smiths Shiprepairers (North Shields) Ltd.* [1984] 1 All E.R. 881; [1984] I.C.R. 236 (where damages were awarded only for increased deafness, etc., occurring after 1963, when employers started to take precautions). For the correct amount of damages for impaired hearing, see *Smith* v. *British Rail Engineering, The Times*, June 27, 1980, C.A. and *Robinson* v. *British Rail Engineering, The Times*, June 27, 1980, C.A.; [1981] 8 C.L. 63. For discovery of documents, etc., in deafness cases, see *Drennan* v. *Brooke Marine Ltd.* and *Kirkup* v. *British Rail Engineering Ltd.*, C.A. [1983] 1 W.L.R. 1165; [1983] 3 All E.R. 147, C.A., and *Edwards* v. *Stroud Riley & Co.* [1984] 7 C.L. 309a (particulars refused against plaintiff).
[73] *Carragher* v. *Singer Manufacturing Co. Ltd.*, 1974 S.L.T. (Notes) 28.
[74] See pp. 101 *et seq.*
[75] *Thompson* v. *Smiths Shiprepairers (N. Shields) Ltd.* [1984] 1 All E.R. 881; [1984] I.C.R. 236.
[76] *Finch* v. *Telegraph Co.* [1949] W.N. 57; [1949] 1 All E.R. 452.

clear directions where he is to get them.[77] In the case of safety spats, it has been held by the House of Lords to be sufficient if the employer, to the knowledge of the employee, has them available in the stores for the asking.[78] There is a difference between a duty of providing something and a duty of seeing that it is used.[79]

(9) NEGLIGENCE OF FELLOW EMPLOYEE

(a) Generally

Since the abolition of the doctrine of common employment (under which an employer was not liable for an accident caused to one employee by a casual act of negligence on the part of another employee) an employer is liable for the negligent acts of his employees, provided that they occur in the course of their employment, which cause injury to another employee.[80] The standard of care required does not differ according to whether the employee is being sued personally or his employer is being sued in respect of his acts or omissions in the course of his employment.[81] The standard of care required in law is that of a reasonably prudent employee.[82]

(b) Practical joking by an employee

An employer ordinarily will not be vicariously liable for injury caused by a practical joke by one employee towards another, for such conduct is outside the course of employment,[83] but he may be liable in respect of a practical joke for failing to take reasonable care to supervise the employee concerned or for failing to provide reasonably competent fellow workmen.[84] The correct test would

---

[77] *Finch* v. *Telegraph Co.*, *supra*, approved by the Court of Appeal in *Clifford* v. *Charles Challen* [1951] 1 K.B. 495; [1951] 1 All E.R. 72, C.A.
[78] *Qualcast (Wolverhampton)* v. *Hayes* [1959] A.C. 743; [1959] 2 All E.R. 38, H.L.
[79] *Norris* v. *Syndic* [1952] 2 Q.B. 135; [1952] 1 All E.R. 935, C.A.
[80] As to vicarious liability generally, see pp. 37–43.
[81] *Staveley Iron and Chemical Co.* v. *Jones* [1956] A.C. 627; [1956] 1 All E.R. 403, H.L.
[82] *Vincent* v. *P.L.A.* [1957] 1 Lloyd's Rep. 103, C.A.
[83] *Smith* v. *Crossley Bros.*, 95 S.J. 655; (1951) C.L.C. 6831, C.A. See also *O'Reilly* v. *National Rail and Tramway Appliances Ltd.* [1966] 1 All E.R. 499; *Coddington* v. *International Harvester Co. of G.B. Ltd.* (1969) 113 S.J. 265; 6 K.I.R. 146 and *Wood* v. *Dutton's Brewery Ltd.* (1971) 115 S.J. 186, C.A. *Cf. Chapman* v. *Oakleigh Animal Products Ltd.* (1970) 114 S.J. 432, C.A.
[84] *Hudson* v. *Ridge Manufacturing Co.* [1957] 2 Q.B. 348; [1957] 2 All E.R. 229.

seem to be whether or not the employer had any reason to suspect that an employee would be likely to perpetrate a practical joke, and whether such practical joke was likely to result in injury.

(10) DEFENCES

(a) Contributory negligence

If an accident is caused partly by the negligence of the employer and partly by the contributory negligence of the plaintiff himself, under section 1(1) of the Law Reform (Contributory Negligence) Act 1945, the plaintiff's damages will be reduced on account of his contributory negligence to such extent as the court thinks just and equitable, provided that the defence of contributory negligence is pleaded by the defendant.[85] Under the *de minimis* principle, an insignificant degree of contributory negligence may be ignored. But it has been held that where a not insignificant degree of contributory negligence is established, the Act does not permit a court to make no reduction at all on the grounds that it considers it just and equitable that there should be no reduction.[86] Moreover, in appropriate cases contribution by the plaintiff can be assessed at 100 per cent., thus negating any award of damages.[87] "Contributory" in this context means "something which is a direct cause of the accident."[88] Unlike negligence, contributory negligence does not depend on the existence of a duty of care.[89] "When contributory negligence is set up as a defence, its existence does not depend on any duty owed by the injured party to the party sued, and all that is necessary to establish such defence is to prove . . . that the injured party did not in his own interest take reasonable care of himself and contributed, by this want of care, to his own injury."[90] Contributory negligence, therefore, may be said to consist of such a failure by an injured party to take reasonable care of

---

[85] If, therefore, the defendant has, by order of the court, been debarred from defending the action, the court cannot, of its own initiative, reduce the plaintiff's damages for contributory negligence—*Fookes* v. *Slaytor, The Times*, June 19, 1978, C.A.

[86] *Boothman* v. *British Northrop Ltd.* (1972) 13 K.I.R. 12, C.A.

[87] *Johnson* v. *Croggan & Co.* [1954] 1 W.L.R. 195; *Cf. Hewson* v. *Grimsby Fishmeal Co.* [1986] 9 C.L. 213 and *Bacon* v. *Jack Tighe (Offshore) and Cape Scaffolding* [1987] 8 C.L. 195.

[88] *Caswell* v. *Powell Duffryn* [1940] A.C. 152, H.L., *per* Lord Porter.

[89] *Jones* v. *Livox* [1952] 2 Q.B. 608; [1952] 1 T.L.R. 1377, C.A.

[90] *Nance* v. *British Columbia Electric Ry.* [1951] A.C. 601; [1951] 2 All E.R. 448, P.C., *per* Lord Simon.

himself as to constitute one of the direct causes of the accident. If there are two or more defendants, the plaintiff's position must be considered separately in regard to each.[91]

The fact that the plaintiff is a minor, *i.e.* aged under 18 years, does not necessarily prevent him being contributorily negligent—it is a question of fact in each case.[92]

What does and does not constitute a failure to take reasonable care in this sense has been the subject of many decisions. Until 1945, contributory negligence in any degree was a complete defence to an action, and there may in consequence have been a judicial tendency to refuse to regard as contributory negligence a minor lapse by an employee who was required to work in conditions of unnecessary danger. Thus, in one frequently quoted dictum, it was said that "it is not for every risky thing which a workman in a factory may do in his familiarity with the machinery that a plaintiff ought to be held guilty of contributory negligence."[93] But in a subsequent decision the House of Lords have emphasised that this dictum must not be given too wide an application:

> I doubt very much whether they (*i.e.*, the words of the dictum) were ever intended to be applied, or could properly be applied, to a simple case of common law negligence, such as the present, where there is no evidence of work people performing repetitive work under strain or for long hours at dangerous machines.[94]

Therefore, although a risky act done by an employee in his familiarity with the operation may not constitute contributory negligence, where there has been a breach of statutory duty by the employer, or where the work being done is of a repetitive nature, the same rule will usually not apply in a case of ordinary negligence at common law.

An employee who tripped on an object on the floor while intent on his job has been held not to be guilty of contributory negligence.[95] Similarly, it was held that there was no contributory negli-

---

[91] *Fitzgerald* v. *Lane* [1987] 2 All E.R. 455, C.A. (Plaintiff and two defendants—damages reduced by one-half, not one-third).
[92] *Minter* v. *D. and H. Contractors (Cambridge)*, The Times, June 30, 1983.
[93] *Flower* v. *Ebbw Vale Steel* [1934] 2 K.B. 132, *per* Lawrence J. at p. 139; approved by Lord Wright in *Caswell* v. *Powell Duffryn* [1940] A.C. 152; [1939] 3 All E.R. 722, H.L., and by Lord Reid in *Summers* v. *Frost* [1955] A.C. 740; [1955] 1 All E.R. 870, H.L.
[94] *Staveley Iron & Chemical Co.* v. *Jones* [1956] A.C. 627; [1956] 1 All E.R. 403 H.L., *per* Lord Tucker.
[95] *Callaghan* v. *Fredd Kidd* [1944] K.B. 560; [1944] 1 All E.R. 525, C.A. *Cf. Burns* v. *British Railways Board* [1977] 10 C.L. (unreported), C.A.

gence where an employee, in a momentary lapse from alertness, fell down the unguarded tank top of a ship. And a workman who slipped on a vertical steel ladder which was not a safe access was not guilty of contributory negligence in failing to place his feet properly on the ladder.[96] An instinctive or semi-instinctive act performed by an employee concentrating on his job will also be judged lightly; thus, where a plaintiff ducked under the guard rail of a platform and inadvertently trod on the trap end of a scaffold he was held only 10 per cent. to blame.[97] But, in general, the question whether inadvertence in itself may amount to negligence is a question of fact which depends on the particular circumstances of each case.[98] The prevailing practice set or allowed by the employer can be an important factor. Thus, where a window cleaner failed to take proper precautions, it was held not to be contributory negligence, as he was only doing the work in the way which the employers expected him to do it.[99]

(b) Volenti non fit injuria (Consent negates legal injury)

The defence shortly expressed in this legal maxim is to the effect that, where a plaintiff with full knowledge of a risk voluntarily accepts and incurs it, this will constitute a complete defence to an action. In practice, however, the maxim can hardly if ever be applied to a case arising between employee and employer, for it cannot be said that an employee voluntarily incurs a risk when he is incurring it in pursuance of the duty which his employer imposes on him.[1]

If, however, an employee undertakes a task which he is not qualified to do, and which the employer has expressly forbidden him to carry out, he cannot at common law claim damages against his employer for injuries received in attempting that task.[2] Simi-

---

[96] *Graham* v. *Scott's Shipbuilding & Engineering Co., Ltd.*, 1963 S.L.T. (Notes) 78.
[97] *Simmons* v. *Bovis* [1956] 1 W.L.R. 381; [1956] 1 All E.R. 736.
[98] *Hicks* v. *B.T.C.* [1958] 1 W.L.R. 493; [1958] 2 All E.R. 39, C.A. *Cf. Ryan* v. *Manbre Sugars Ltd.* (1970) 114 S.J. 492, C.A., where it was stated that inadvertence is not contributory negligence.
[99] *General Cleaning Contractors* v. *Christmas* [1953] A.C. 180; [1952] 2 All E.R. 1110, H.L.
[1] *D'Urso* v. *Sanson* [1939] 4 All E.R. 26.
[2] *Hyland* v. *R.T.Z. (Barium Chemicals)* [1975] I.C.R. 54 (productivity agreement permitting flexibility did not alter the position). *Cf. Uddin* v. *Associated Portland Cement Manufacturers* [1965] 2 Q.B. 582, C.A. (breach by employers of statutory duty: employee injured in factory while on "frolic of his own"—employer liable, but subject to 80 per cent. reduction of damages for employee's contributory negligence).

larly, if an employee is in breach of a statutory duty imposed on him personally, he may be held to have consented to consequent injury, even though another employee is involved.[3]

It is provided by section 2 of the Unfair Contract Terms Act 1977[4-5] that, "where a contract term or notice purports to exclude or restrict liability for negligence a person's agreement to or awareness of it is not of itself to be taken as indicating his voluntary acceptance of any risk." This further fetters an employer's ability to raise the defence of *volenti non fit injuria* against his employee.

(11) LIABILITY FOR ACTS OF OTHERS

(a) Generally

An employer, in common with other defendants, is liable for the acts of his servants acting in the course of their employment and for the acts of his agents acting within the scope of their authority. An employer, also, will ordinarily be liable for the acts of an independent contractor whom he has appointed to fulfil some part of the duty of care which he owes to his employees. Where the employer tells his employee to obey the orders of a third person or otherwise places his employee under that person's control, that person may well become the employer's employee *ad hoc* and the employer is vicariously liable to his employee for injuries sustained as a result of the third person's negligence.[6] If the employer is made liable for the act of another (vicarious liability) he is able to claim from that other such contribution (which can be 100 per cent. indemnity) "as may be found by the court to be just and equitable having regard to the extent of that person's responsi-

---

[3] *Imperial Chemical Industries Ltd.* v. *Shatwell* [1965] A.C. 656, H.L.; *McMullen* v. *National Coal Board* [1982] I.C.R. 148.

[4-5] 1977 (c. 50). The corresponding provision in the 1977 Act for Scotland (s.16(3)) refers only to a term of a contract and does not include a notice. It is therefore in theory still possible for an employer in Scotland to invoke the defence of *volenti* by a notice at the place of work but the employer would still have to overcome the rule that *sciens* is not *volens* (knowledge is not *per se* consent)—see *Smith* v. *Baker & Sons* [1891] A.C. 325, H.L.

[6] *McDermid* v. *Nash Dredging and Reclamation Co. Ltd.* [1987] 2 All E.R. 878; *The Times*, July 3, 1987, H.L.—deck-hand had to obey orders of tug master (not otherwise employee of defendants). *Cf. Kondis* v. *State Transport Authority* (1984) 154 C.L.R. 672, High Ct. of Australia; and *Ferguson* v. *Welsh and Others* [1987] 3 All E.R. 777, H.L. (Occupier of building site not liable, *qua* employer, to unauthorised sub-contractor's employee injured by unsafe system of working).

bility for the damage in question."⁷ The employer may also be able to recover an indemnity under an express or implied term of the contract.⁸ Claims for contribution or indemnity, of all kinds, survive the death of the party claiming and pass to his executors and administrators, even if at the death the claim had not been established.⁹

(b) Course of employment

An employer is not liable for the acts of his employee committed outside the course of the employment of the latter.¹⁰ Where an employee suggested to the plaintiff that he should load cargo on to the hatch cover of a ship, and the hatch cover gave way, thereby injuring the plaintiff, it was held that the employer of the employee was not liable, as it was neither in the course of his employment nor within the scope of his authority to make such a suggestion.¹¹ But an act which is expressly prohibited by the employer may still be within the course of employment, for the test is not whether the manner of doing the act was authorised or not, but whether the act itself was authorised; thus, where an employee was instructed to make a journey using a vehicle of the employers, and instead he took his own car for the purpose, it was held that he was acting within the course of his employment, for the act in this case was the making of the journey itself, and the fact that the journey had been performed in an unauthorised manner was irrelevant.¹² Similarly, where a servant made a journey by

---

⁷ Civil Liability (Contribution) Act 1978 (c. 47), ss.1 and 2, replacing the Law Reform (Married Women and Tortfeasors) Act 1935, s.6, as to which see *Forte's Service Areas Ltd.* v. *Department of Transport, The Times*, July 31, 1984, C.A. (contribution obtainable from any other tortfeasor if he would have been liable if sued at any time). For the position where the employer has settled the claim, see s.3 of the 1978 Act; *Logan* v. *Uttlesford District Council, The Times*, February 21, 1984, and *Harper* v. *Gray-Walker* [1985] 2 All E.R. 507. The 1978 Act is not confined to liabilities incurred in England and Wales, see *Virgo Steamship Co. SA* v. *Skaarup Shipping Corp. and Others, The Times*, October 21, 1987.
⁸ See *Lister* v. *Romford Ice Storage Ltd.* [1957] A.C. 555, H.L., but *cf. Harvey* v. *O'Dell* [1958] 2 Q.B. 78; *Morris* v. *Ford Motor Co. Ltd.* [1973] 2 All E.R. 1084, C.A., and *Southern Water Authority* v. *Carey* [1985] 2 All E.R. 1077.
⁹ *Ronex Properties Ltd.* v. *John Laing Ltd., The Times*, July 28, 1982, C.A.
¹⁰ As to the effect of port by-laws on the vicarious liability of the port authority, see *Karuppan Bhoomidas* v. *Port of Singapore Authority* [1978] 1 All E.R. 956; (1977) 121 S.J. 816, P.C.
¹¹ *Hillen* v. *I.C.I.* [1936] A.C. 65, H.L.
¹² *McKean* v. *Raynor Bros.* [1942] 2 All E.R. 650, applied in *Harvey* v. *O'Dell* [1958] 2 Q.B. 78; [1958] 1 All E.R. 657. See also *Ilkiw* v. *Samuels* [1963] 1 W.L.R. 991; [1963] 2 All E.R. 879, C.A., and *Angus* v. *Glasgow Corp.*, 1977 S.L.T. 206 (Scot.).

car instead of going by train as instructed, it was held that he was within the course of his employment.[13] The general principle has been stated as follows:

> "It is well-settled law that a master is liable even for acts which he has not authorised, provided that they are so connected with the acts which he has authorised that they may rightly be regarded as modes, although improper modes, of doing them. On the other hand, if the unauthorised and wrongful act of the servant is not so connected with the authorised act as to be a mode of doing it, but is an independent act, the master is not responsible, for in such a case the servant is not acting in the course of his employment but has gone outside it."[14]

Thus a driver who is forbidden to give lifts will be outside the course of his employment if he does so.[15] However, the evidence may show an express or implied authority to an employee by his employer to transport employees to and from work, either in the employer's or the employee's vehicle, in which case the employee is vicariously liable for the employee's negligent driving.[16] A similar result was reached where an apprentice, who had no driving licence and was not an authorised driver, drove his employer's van in a workshop in order to clear a space for certain equipment to be moved in, and in so doing fatally injured another workman.[17] But where a bus conductor, who was prohibited from driving a bus in any circumstances, drove a bus in order to move it to enable his driver to take his bus out of the garage, it was held that his employers were not liable for the accident which resulted. The prohibition did not merely relate to the conduct of his employ-

---

[13] *Canadian Pacific Ry.* v. *Lockhart* [1942] A.C. 591; [1942] 2 All E.R. 464, P.C.
[14] *Marsh* v. *Moores* [1949] 2 K.B. 208; [1949] 2 All E.R. 27, *per* Lynskey J. See also *Kay* v. *I.T.W.* [1968] 1 Q.B. 140, C.A.; *East* v. *Beavis Transport Ltd.* [1969] 1 Lloyd's Rep. 302, C.A.
[15] *Twine* v. *Bean's Express* (1946) 175 L.T. 131; 62 T.L.R. 458, C.A.; *Conway* v. *George Wimpey* [1951] 2 K.B 266; [1951] 1 All E.R. 363, C.A., distinguished in *Rose* v. *Plenty* [1976] 1 All E.R. 97; (1975) 119 S.J. 592, C.A. (injured boy invited, in breach of prohibition, to help on milkround: milkman's employers liable). The employer may not, however, be liable criminally in such circumstances—see *Portsea Island Mutual Co-operative Society Ltd.* v. *Leyland* [1978] I.C.R. 1195, D.C.
[16] *Cf. Canadian Pacific Ry. Co.* v. *Lockhart* [1942] A.C. 591, P.C.; *McKean* v. *Raynor Bros. Ltd. (Nottingham)* [1942] 2 All E.R. 650; *Semtex Ltd.* v. *Gladstone* [1954] 2 All E.R. 206; *Harvey* v. *R.G. O'Dell Ltd.* [1958] 1 All E.R. 657; and *Paterson* v. *Costain and Press (Overseas) Ltd.* [1978] 1 Lloyd's Rep. 86.
[17] *Mulholland* v. *William Reid & Leys*, 1958 S.C. 290; 1958 S.L.T. 285.

ment, but placed driving a bus outside the sphere of his employment.[18]

Where an employee assaults a customer,[19] or where an ordinary miner fires a shot in a mine,[20] or where a lorry driver walks across a road from his lorry in order to get refreshment at a transport café, and in so doing causes an accident,[21] it has been held not to be in the course of employment.

On the other hand, the mere fact that an act is a criminal act will not necessarily mean that it is outside the course of employment. Thus a lorry driver who stole the goods of a consignor in transit was held to be acting in the course of employment,[22] and if an act is properly part of or fairly incidental to the day's work, such as where an employee on an outside job made a journey to get tools and a meal, it will be within the course of employment.[23] In this connection, it has been held that an employee who was bicycling across a bus park in factory premises to draw his wages was acting in the course of his employment, but probably would not have been if he had not been on the premises at the time,[24] while a sailor returning to his ship from shore leave in a hired dinghy has been held not to be acting in the course of his employment.[25] Travel to and from the place of employment is only within the course of employment where the employee is contractually obliged to travel in the transport provided by the employer.[26] The provision in section 53 of the Social Security Act 1975 that an employee is deemed to be in the course of his employment even though not under any obligation to travel in any vehicle, only alters the law for the special purposes of that Act. It does not affect the established interpretation of "course of employment" in the Workmen's Compensation Acts, the Road Traffic Acts, employer's liability insur-

---

[18] *Iqbal* v. *London Transport Executive, The Times*, June 7, 1973, C.A.
[19] *Warren* v. *Henlys* [1948] W.N. 449; [1948] 2 All E.R. 935. *Cf.* assault on fellow-employee—employer may be liable—*Duffy* v. *Thanet District Council* (1984) 134 New L.J. 680.
[20] *Alford* v. *N.C.B* [1952] W.N. 144; [1952] 1 All E.R. 754, H.L.
[21] *Crook* v. *Derbyshire Stone* [1956] 1 W.L.R. 432; [1956] 2 All E.R. 447.
[22] *United Africa Co.* v. *Saka Owoade* [1955] A.C. 130; [1957] 3 All E.R. 216, P.C.
[23] *Harvey* v. *O'Dell* [1958] 1 All E.R. 657.
[24] *Staton* v. *N.C.B.* [1957] 1 W.L.R. 893; [1957] 2 All E.R. 667; *Cf. Compton* v. *McClure* [1975] I.C.R. 378.
[25] *Bradford* v. *Boston Deep Sea Fisheries* [1959] 1 Lloyd's Rep. 394. *Cf. Paterson* v. *Costain and Press (Overseas) Ltd.* (1979) 123 S.J. 142, C.A.—Injury to employee when passenger in employers' vehicle, returning from leave to construction site in Iran—*held* "in course of employment."
[26] Firemen proceeding to a fire station to answer to an alarm call are a possible exception to this principle: *Stitt* v. *Woolley* (1971) 115 S.J. 708, C.A.

ance policies, and for the purposes of vicarious liability generally.[27]

Where an employer owes a personal duty of care, whether to his employee[28] or to a third party, *e.g.* to a bailor to take care of his goods,[29] a breach of that duty caused by a servant of the employer acting on a "frolic of his own" will nevertheless impose liability on the employer to the employee or the third party, as the case may be.

(c) Loan of servant

Where an employee who is on loan from a general to a temporary employer is negligent, the question may well arise as to which of the two alleged employers is vicariously liable for such negligence. The similar position where the servant on loan is himself injured has already been considered,[30] and although the broad principles are the same in both cases, the test to be applied may well differ.[31]

In cases where the loaned servant is himself negligent, the test of vicarious liability turns on where the authority lies to direct, or to delegate to, the workman the manner in which the work is to be performed.[32] "The ultimate question is not what specific orders, or whether any specific orders, were given, but who is entitled to give the orders as to how the work should be done."[33] Generally speaking, drivers of vehicles and cranes will remain the servants of their

---

[27] *Vandyke* v. *Fender* [1970] 2 Q.B. 292; [1970] 2 W.L.R. 929, C.A.; reversing [1969] 2 Q.B. 581 on this point. See also *Nottingham* v. *Aldridge* [1971] 2 Q.B. 739.
[28] See, *e.g. Hudson* v. *Ridge Manufacturing Ltd.* [1957] 2 Q.B. 348 (one employee injured by practical joke of another, known to employer to be addicted to joking, employer liable for breach of personal duty to provide safe fellow–employees).
[29] *Morris* v. *C.W. Martin Ltd.* [1966] 1 Q.B. 716; [1965] 2 All E.R. 725, C.A. *Cf. Photo Production Ltd.* v. *Securicor Transport Ltd.* [1978] 3 All E.R. 146, C.A. (Securicor liable when their employee deliberately set on fire a factory that Securicor had contracted to guard), reversed by the House of Lords on another point but not affecting the above ruling [1980] 1 All E.R. 556, H.L. Moreover, whatever the nature of the bailment, the onus is upon the bailee to prove that the loss of the goods bailed to him was not caused by any fault of his or of his servants or agents to whom he entrusted the goods for safekeeping—*Port Swettenham Authority* v. *T.W. Wu & Co.* [1978] 3 All E.R. 337, P.C.
[30] See pp. 6–7.
[31] *Garrard* v. *Southey (F.E.)* [1952] 2 Q.B. 174; [1952] 1 All E.R. 597.
[32] *Mersey Docks and Harbour Board* v. *Coggins & Griffith* [1947] A.C. 1; [1946] 2 All E.R. 345, H.L., *per* Lord Simon.
[33] *Mersey Docks and Harbour Board* v. *Coggins & Griffith, supra, per* Lord Porter, applied by the House of Lords in *John Young & Co. (Kelvinhaugh) Ltd.* v. *O'-Donnell*, 1958 S.L.T. (Notes) 46.

general employers, even though the agreement of hiring may provide otherwise.[34]

(d) Independent contractors[35]

The basic principle is that an employer who employs an independent contractor is not liable for torts committed by that contractor or employees of the contractor. "Exceptions" to this principle exist where the employer owes a personal duty to the person injured which he cannot delegate to a contractor to perform on his behalf. Since, in such circumstances, the employer is himself in breach of duty, they are not true exceptions to the basic principle.

Clearly, an employer may be liable if he negligently selects an incompetent contractor. There are various situations when an employer may be liable where he commissions an independent contractor to perform what have been described as "extra-hazardous or dangerous operations."[36]

When an independent contractor is employed to discharge a non-delegable statutory duty, the employer will be liable for the independent contractor's failure to perform that duty. Whether or not a statutory duty is non-delegable is a question of construction. The duties imposed by section 14 and section 29(1)[37] of the Factories Act 1961 (secure guarding of dangerous machinery and safe means of access, etc., respectively) are non-delegable.[38] In all these circumstances outlined so far, the employer's duty is sometimes described as one to see that care is taken rather than just a duty to take reasonable care.[39]

With respect to an employer's duty to take reasonable care for the safety of his employees, it appears to be established that he cannot perform that duty simply by appointing competent contrac-

---

[34] *Willard* v. *Whiteley* [1938] 3 All E.R. 779, C.A.; *Century Insurance* v. *Northern Ireland R.T.B.* [1942] A.C. 509; [1942] 1 All E.R. 491, H.L.; *Bontex* v. *St. John's* [1944] 1 All E.R. 381n., C.A.; *Mersey Docks and Harbour Board* v. *Coggins & Griffith, supra*; *John Young & Co. (Kelvinhaugh) Ltd.* v. *O'Donnell, supra.*

[35] The scope of an employer's liability for the acts of independent contractors employed by him was extensively considered in *Salsbury* v. *Woodland* [1970] 1 Q.B. 324, C.A.; *cf. Maguire* v. *P.J. Lagan (Contractors)* [1976] N.I. 49, C.A.

[36] *Honeywill & Stein Ltd.* v. *Larkin Bros. Ltd.* [1934] 1 K.B. 191 *per* Slesser L.J. at 200.

[37] *Hosking* v. *De Havilland Aircraft Co. Ltd.* [1949] 1 All E.R. 540.

[38] *Cf.* the duties imposed by the Construction (Working Places) Regulations 1966. See *Donaghey* v. *Boulton and Paul Ltd.* [1968] A.C. 1.

[39] *The Pass of Ballater* [1942] P. 112 at 117.

tors to do the work.[40] Thus, for example, an employer will be liable to an employee injured because of negligent installation of plant by independent contractors.[41]

(12) DAMAGES

(a) Object of damages

The object of an award of damages in tort generally is to compensate the plaintiff for the loss he has suffered. This loss is measured, broadly speaking, by contrasting the position of the plaintiff before and after the commission of the tort. With respect to damages for personal injuries, the chief difficulty in assessing damages is quantifying in financial terms the non-pecuniary loss such injuries involve. As an aid to quantification of such damages at an early stage, there are statutory powers[42] for the High Court to order disclosure to a party[43] or intended party[44] to personal injuries litigation of relevant documents,[45] *e.g.* medical records unless compliance with the order would be likely to be injurious to the public interest.[46] Where the action has been started, the order

---

[40] *Paine* v. *Colne Valley Electricity Supply Co.* [1938] 4 All E.R. 803, *per* Goddard L.J. at 807; *Cassidy* v. *Ministry of Health* [1951] 2 K.B. 343, *per* Denning L.J. at 363; approved by Morris L.J. in *Walsh* v. *Holst & Co.* [1958] 1 W.L.R. 800 at 806–807; [1958] 3 All E.R. 33 at 38; *Green* v. *Fibreglass* [1958] 2 Q.B. 245, *per* Salmon J. at 250.
[41] *Paine* v. *Colne Valley Electricity Supply Co.* [1938] 4 All E.R. 803.
[42] Under the Administration of Justice Act 1970 (c. 31), ss.31–35 (passed to implement the recommendations of the Winn Committee on Personal Injuries Litigation, Cmnd. 3691). This is a wide power and it does not have to be shown that the applicant would find it impossible to conduct his case without the documents—*O'Sullivan* v. *Herdmans Ltd.*, *The Times*, July 10, 1987, H.L.
[43] Disclosure must be to the party (who may elect to have disclosure to his agent) and the court cannot restrict disclosure, *e.g.* to the party's medical adviser—*McIvor and Another* v. *Southern Health and Social Services Board, Northern Ireland* [1978] 2 All E.R. 625, H.L.
[44] "Documents" include X-ray photographs—*McCarthy* v. *O'Flynn* [1979] I.R. 127, Eire Sup.Ct.
[45] If the dominant purpose for which the document was prepared was in order to obtain legal advice, or for the purpose of litigation, the document is covered by legal privilege and need not be disclosed—*Waugh* v. *British Railways Board* [1979] 3 W.L.R. 150; [1979] 2 All E.R. 1169, H.L. (report of internal inquiry on railway accident, prepared for dual purpose, not privileged); *cf. Lee* v. *S.W. Thames Regional Health Authority* [1985] 1 W.L.R. 845, C.A.; *Farrell* v. *Rotary-Services* [1985] 1 N.I.J.B. 55; *Earle* v. *Medhurst* [1985] 12 C.L. 296.
[46] See *O'Sullivan* v. *Herdmans Ltd.* [1987] 1 W.L.R. 1047, H.L. (production order of DHSS documents relating to prescribed disease claim).

can be made against a person not a party to the action, *e.g.* a doctor or a hospital holding medical records or the Department of Health and Social Security.[47] A plaintiff cannot normally insist on having his doctor present at an examination of the plaintiff by the defendant's medical adviser.[48] Nor can he insist as a precondition to examination on seeing the whole of the defendant's medical report.[49]

It is possible, under the Rules of the Supreme Court,[50] for a plaintiff in a personal injuries action to lessen some of these difficulties by applying for an interim payment of damages. This is only possible in certain cases, detailed in the Rules. A similar procedure is applicable in Scotland.[51] A new procedure for the award of provisional damages where there is a chance of serious deterioration in the plaintiff's condition is introduced by section 6 of the Administration of Justice Act 1982.

(b) Remoteness of damage

Whether a particular loss that results from the defendant's tortious act is too remote a consequence of that act to permit recovery for it is a problem closely related to whether the defendant's conduct is prima facie tortious in the first place. With respect to damages for personal injuries, the problem of remoteness may arise where the plaintiff suffers unusual or unexpected injuries. The correct test is whether or not the damage that occurred was reasonably foreseeable.[52] But it is sufficient if the plaintiff establishes that the type of damage that occurred was reasonably foreseeable, even though

---

[47] *Hall* v. *Avon Area Health Authority (Teaching)* [1980] 1 All E.R. 516, C.A. As to the circumstances in which a plaintiff can refuse to submit to a particular type of medical examination on behalf of the defendant, *e.g.* on the ground that it is painful or risky, see *Starr* v. *National Coal Board* [1977] 1 All E.R. 243, C.A.; *Aspinall* v. *Sterling Mansell Ltd.* [1981] 3 All E.R. 866; and *Prescott* v. *Bulldog Tools Ltd.* [1981] 3 All E.R. 869.
[48] *Megarity* v. *D.J. Ryan & Sons* [1980] 2 All E.R. 832, C.A.
[49] [1981] 1 W.L.R. 549; [1981] 2 All E.R. 21, C.A.
[50] R.S.C., Ord. 29, r. 9.
[51] See *Douglas C.B.* v. *Douglas and Another*, 1974 S.L.T. (Notes) 67; *Littlejohn* v. *Clancy*, 1974 S.L.T. (Notes) 68; *Boyle* v. *Rennies of Dunfermline Ltd.*, 1975 S.L.T. (Notes) 13; *Nelson* v. *Duraplex Industries Ltd.*, 1975 S.L.T. (Notes) 31.
[52] *The Wagon Mound* [1961] A.C. 388, P.C.; For the application of this test to the acts of third parties intervening to cause, or increase, the damage—see *Lamb* v. *Camden London Borough Council* [1981] 2 W.L.R. 1038; [1981] 2 All E.R. 408, C.A.

the precise way in which it came about was not.[53] Further, if the type of damage was reasonably foreseeable the plaintiff can recover for the full extent of his injuries even though they may be more extensive than would normally be expected due to his own physical condition.[54]

(c) Quantification of damage

The loss that follows an accident causing personal injuries can be divided into two categories: financial and non-financial. The financial loss may be of two types: loss of earnings and medical expenses. Loss of earnings incurred between the date of the accident and the trial may be recovered as special damages. Where the accident diminishes the plaintiff's future earning capacity, he can also recover a sum in respect of loss of future earnings as part of an award of general damages.

He may also recover damages for loss of benefits under an occupational pension scheme but not for the value of contributions he would have made to the pension scheme out of the lost earnings.[55] Where the accident also shortens the victim's life expectation the period for which he may recover for loss of future earnings is not limited to the time for which he is now likely to live but will be the entire period of his life expectation before the accident, etc. (though the damages must be reduced for his probable living expenses during that period).[56] Medical expenses are similarly divided into those incurred before the trial of the action, which are special damages, and the estimated cost of medical treatment and nursing after the trial, which may be taken into account in awarding general damages. While it is true that the plaintiff may receive all his treatment under the National Health Service and thus suffer

---

[53] *Hughes* v. *Lord Advocate* [1963] A.C. 837; 1963 S.C., H.L. 31; 1963 S.L.T. 150, H.L.; *Bradford* v. *Robinson Rentals Ltd.* [1967] 1 W.L.R. 337; [1967] 1 All E.R. 267; *cf. Doughty* v. *Turner Manufacturing Co. Ltd.* [1964] 1 Q.B. 518, C.A., and *Holland* v. *British Steel Corp.*, 1974 S.L.T. (Notes) 72.
[54] *Smith* v. *Leech Brain & Co. Ltd.* [1962] 2 Q.B. 405; *Warren* v. *Scruttons* [1962] 1 Lloyd's Rep. 497; *Robinson* v. *Post Office* [1974] 2 All E.R. 737, C.A.; *cf. Tremain* v. *Pike* [1969] 1 W.L.R. 1556; [1969] 3 All E.R. 1303. This principle applies equally to mental conditions: *Malcolm* v. *Broadhurst* [1970] 3 All E.R. 508.
[55] *Dews* v. *National Coal Board* [1987] 3 W.L.R. 38, H.L.
[56] *Pickett* v. *British Rail Engineering Ltd.* [1980] A.C. 136; [1979] 1 All E.R. 774, H.L.; *Lim* v. *Camden and Islington Area Health Authority* [1979] 2 All E.R. 910, H.L. *Contra* where the plaintiff is a young child, when damages for loss of earnings (though recoverable) are confined to the actual, reduced, life expectation— *Croke* v. *Wiseman* [1981] 3 All E.R. 853, C.A. See also s.1(2) of the Administration of Justice Act 1982 (c. 53).

no loss in this respect, the courts will consider a claim under this head if the plaintiff has received or wishes to receive private treatment.[57] Similarly, if a severely injured plaintiff proves that it is reasonable for him to be cared for at home, he is entitled to recover damages for the future cost of such care, even though greater than would be incurred if he were in a private institution.[58]

Non-financial loss may be considered under three heads which may result from physical or mental injury or from work-related disease.[59] There is no fixed upper limit for damages for pain, suffering and loss of amenity.[60] The first, pain and suffering, includes knowledge of impending death and nervous shock[61] as well as physical pain and suffering caused by the accident. If the accident renders the victim unconscious, he cannot recover damages under this head.[62] The second head, loss of amenities, covers not only loss of limbs and faculties but also the lessening of the victim's enjoyment of life by restricting or preventing his former activi-

---

[57] *Oliver* v. *Ashman* [1962] 2 Q.B. 210, C.A. The victim of an accident causing personal injuries has a duty to mitigate his loss by seeking medical attention. If he unreasonably refuses to avail himself of medical attention, and his refusal becomes the proximate cause of his incapacity, it can no longer be said to result from the accident: *Steel* v. *Robert George & Co. Ltd.* [1942] A.C. 497, *per* Lord Simon. The reasonableness of his refusal has to be judged by the objective standard of the reasonable man: *Morgan* v. *T. Wallis Ltd.* [1974] 1 Lloyd's Rep. 165. If the plaintiff is being maintained at public expense in, *e.g.* a hospital, the saving in living expenses must be deducted from any claim for loss of income—Administration of Justice Act 1982 (c. 53), s.5.
[58] *Rialas* v. *Mitchell, The Times*, July 17, 1984, C.A. *Cf. Housecroft* v. *Burnett* [1986] 1 All E.R. 332; *The Times*, June 7, 1985, C.A. ("reasonable recompense" to caring relative).
[59] For provisions as to disclosure of medical records, etc., see the statutory powers of the High Court to order disclosure (Administration of Justice Act 1970 (c. 31), ss.31–35.)
[60] *Mustart* v. *Post Office, The Times*, February 11, 1982 (£65,000 damages for loss of sight, taste and smell), *cf. Croke* v. *Wiseman* [1981] 1 W.L.R. 71, C.A., and *Housecroft* v. *Burnett* [1986] 1 All E.R. 332; *The Times*, June 7, 1985, C.A. (£75,000 damages guideline for pain, suffering and loss of amenity in average cases of tetraplegia).
[61] Administration of Justice Act 1982 (c. 53), s.1(1). This covers any recognisable psychiatric illness: *Hinz* v. *Berry* [1970] 2 Q.B. 40, C.A.; *Carlin* v. *Helical Bar Ltd.* (1970) 9 K.I.R. 154; and *Galt* v. *British Railways Board* (1983) 133 New L.J. 870 (shock caused subsequent myocardial infarction). *Cf. Pickett* v. *British Rail Engineering Ltd.* (1977) 121 S.J. 814; *The Times*, November 18, 1977, C.A. (£3,000 damages for distress suffered by plaintiff because of realisation that his dependants would be left without him to care for them), and *Rourke* v. *Barton, The Times*, June 23, 1982 (additional damages for injured wife's knowledge that she could not look after her terminally ill husband).
[62] *Wise* v. *Kaye* [1962] 1 Q.B. 638, C.A.; *H. West & Son Ltd.* v. *Shephard* [1964] A.C. 326, H.L.

ties—e.g. the loss of ability to marry and have children,[63] to play sports and indulge in recreational activities generally. Recovery is possible under this head even though the plaintiff is so seriously injured that he cannot appreciate his loss.[64] If the plaintiff's injuries, e.g. to the head, cause him to be imprisoned for consequent crimes, he can recover additional damages for that. However, he cannot recover the financial loss arising from provision for his wife, etc., he makes in matrimonial proceedings resulting from, e.g. brain damage, though costs of those proceedings are recoverable as damages from the defendant.[65] Damages may no longer be recovered for loss of expectation of life as such, though damages are still recoverable for pain and suffering caused by realisation of diminished expectancy and for loss of income for the "lost" years.[66]

(d) Events after the accident

In assessing damages for personal injuries, a significant degree of speculation may be involved, e.g. as to the plaintiff's future earnings, the long-term physical disabilities he will be under or even his life expectancy.[67] Events may occur after the accident but before the trial or even after the trial but before an appeal[68] which may materially affect one or more of the factors about which the court is required to speculate. Broadly speaking, the court will take

---

[63] *Guasto* v. *Robinsons of Winchester Ltd.*, *The Times*, November 23, 1977 (£11,000 damages for 50 per cent. loss of sexual function). *Cf. Jones* v. *Jones* [1983] 1 All E.R. 1037 (damages can include loss arising out of break-up of plaintiff's marriage, if consequent on his injuries, but only to the extent that the break-up causes increased costs of maintaining his former wife and the children).

[64] *H. West & Son Ltd.* v. *Shephard* [1964] A.C. 326, H.L. *Cf. Croke* v. *Wiseman* [1981] 3 All E.R. 853, C.A.

[65] For additional damages for imprisonment, see *Meah* v. *McCreamer*, *The Times*, July 5, 1984. For matrimonial losses, see *Pritchard* v. *J.H. Cobden Ltd. and Another*, *The Times*, August 27, 1986 and December 3, 1986, C.A. ('over-ruling' *Jones* v. *Jones* [1985] Q.B. 794, C.A.).

[66] Administration of Justice Act 1982 (c. 53), s.1.

[67] For provisions as to disclosure of medical records, etc., see the statutory powers of the High Court to order disclosure (Administration of Justice Act 1970 (c. 31), ss.31–35.)

[68] See *Curwen* v. *James* [1963] 1 W.L.R. 748; [1963] 2 All E.R. 619, C.A.; *Jenkins* v. *Richard Thomas & Baldwins Ltd.* [1966] 1 W.L.R. 476; [1966] 2 All E.R. 15, C.A.; and *Murphy* v. *Stone Wallwork (Charlton) Ltd.* [1969] 1 W.L.R. 1023; [1969] 2 All E.R. 949, H.L., where the case proceeded on the basis that the plaintiff would not lose his job, but he was dismissed after the Court of Appeal decision was given. The House of Lords ordered a review of the general damages awarded in the light of this event. As to when the Court of Appeal may exercise its discretion to admit new evidence, see *Mulholland* v. *Mitchell* [1971] A.C. 666, H.L.

account of these events. Where there is a chance that in the future the injured person will as a result of the wrongful act develop some serious disease or suffer some serious deterioration in his physical or mental condition, the court may, under section 6 of the Administration of Justice Act 1982 make a provisional award of damages, leaving open the possibility of a further award later.

A subsequent accident may exacerbate the extent of the plaintiff's injury. Where he suffers a second accident because of a weakness caused by the original accident the person whose negligence caused the first accident may be liable for the loss flowing from the second unless the plaintiff's conduct was unreasonable in view of the disability he was under following the first accident.[69]

If after an accident, the plaintiff becomes disabled by a latent disease which he had before the accident and is not brought on by the accident, the defendant is not liable to pay damages for loss of earnings for a period beyond the date when the plaintiff becomes disabled.[70]

The normal duty of a plaintiff to take all reasonable steps necessary to mitigate his damage applies to personal injuries and may require a plaintiff to undergo surgery or other medical treatment.[71] If he does not, his damages will be diminished accordingly.

(e) Taxation, social security,[72] etc., and damages

The House of Lords decision in *British Transport Commission* v. *Gourley*[73] established the following principle: in assessing damages not in themselves subject to tax,[74] for loss of income which if received would have been taxable in the recipient's hands, there will be deducted a sum broadly equivalent to the estimated

---

[69] *McKew* v. *Holland and Hannen & Cubitts (Scotland) Ltd.*, 1970 S.C., H.L. 20; [1969] 3 All E.R. 1621, H.L.; *Wieland* v. *Cyril Lord Carpets* [1969] 3 All E.R. 1006. See also *Wilkins* v. *William Cory & Son* [1959] 2 Lloyd's Rep. 98. *Cf. Nolan* v. *Dental Manufacturing Co.* [1958] 1 W.L.R. 936; [1958] 2 All E.R. 449.

[70] *Jobling* v. *Associated Dairies Ltd.* [1981] 2 All E.R. 752, H.L. *Cf. Cutler* v. *Vauxhall Motors Ltd.* [1971] 1 Q.B. 418, C.A. and *Baker* v. *Willoughby* [1970] A.C. 467, H.L.

[71] *Selvanayagam* v. *University of the West Indies* [1983] 1 All E.R. 824, P.C.

[72] For claims by employees under the social security legislation generally see Chap. 7.

[73] [1956] A.C. 185, H.L.

[74] Where however the damages will be assessable to tax by the Inland Revenue (because of the provisions of s.187 of the Income and Corporation Taxes Act 1970 only first £25,000 exempt from tax), damages will be increased to a figure which after the tax has been paid, will represent the plaintiff's actual loss of income—*Shove* v. *Downs Surgical plc* [1984] 1 All E.R. 7.

liability to tax which the recipient would have incurred, had he received the lost income.[75] In an action for damages for personal injuries, therefore, future earnings lost are assessed net of the income tax that would have been paid on those earnings.

The principle has been extended in two ways. First, damages for loss of earnings in respect of any period during which the plaintiff's incapacity prevents him from working at all, must also be assessed net of the National Insurance contributions that would have been paid out of those earnings.[76] Second, credit must be given for certain social security benefits received consequent on the accident. Credit must be given for all of any unemployment benefit received.[77] The principle here is that claiming unemployment benefit is a means of mitigating loss of earnings similar to obtaining another job.[78] The same applies to payments received under the Job Release Scheme[79] and to statutory sick pay under the Social Security and Housing Benefits Act 1982.[80] Similarly, industrial rehabilitation allowance[81] and supplementary benefit[82] are deductible, but attendance or mobility allowances are not deductible, as they are for specific services.[83] So far as concerns industrial injury and disablement benefits (except constant attendance allowance), and sickness benefit, one-half of such benefits received (or probably to be received) in the five years after the injury must, by statute, be deducted from damages for loss of earnings or profits,[84] but no credit need be given for such benefits after the five years'

---

[75] See Law Reform Committee for Scotland, 6th Report 1959, Cmnd. 635, para. 2 cf. *Attree* v. *Baker*, *The Times*, November 18, 1983.

[76] *Cooper* v. *Firth Brown Ltd.* [1963] 1 W.L.R. 418; [1963] 2 All E.R. 31.

[77] *Parsons* v. *B.N.M. Laboratories* [1964] 1 Q.B. 95, C.A. However, in *Nabi* v. *British Leyland (U.K.) Ltd.* [1980] 1 All E.R. 667, the Court of Appeal indicated that this rule ought possibly to be reviewed by the legislature or by the House of Lords.

[78] *Crawley* v. *Mercer*, *The Times*, March 4, 1984.

[79] *Ibid.*

[80] *Palfrey* v. *Greater London Council* [1985] I.C.R. 437.

[81] *Cackett* v. *Earl*, *The Times*, October 19, 1976; [1976] C.L.Y. 666; *Sanger* v. *Kent and Callow* [1978] 11 C.L. 75; *Plummer* v. *Wilkins and Son Ltd.*, *The Times*, July 9, 1980.

[82] *Lincoln* v. *Hayman and Another*, *The Times*, February 18, 1982, C.A.

[83] *Bowker* v. *Rose*, *The Times*, February 3, 1978, C.A.

[84] Law Reform (Personal Injuries) Act 1948 (c. 41), s.2(1) and (2). See *Perez* v. *C.A.V. Ltd.* [1959] 2 All E.R. 414; *Hultquist* v. *Universal Pattern and Precision Engineering Co. Ltd.* [1960] 2 All E.R. 266, C.A.; *Rudy* v. *Tay Textiles* (O.H.) 1978 S.L.T. (Notes) 62; *Lim Poh Choo* v. *Camden and Islington Area Health Authority* [1980] A.C. 174, H.L.; *Bowers* v. *Strathclyde Regional Council* 1981 S.L.T. 122 and *Foster* v. *Tyne and Wear County Council* [1986] 1 All E.R. 567, C.A. (includes award for loss of earning capacity).

period.[85] Payments received under a private insurance policy effected by the employee are similarly ignored[86] but payments to the employee by the employer, reimbursed to the employer by *his* insurers are deductible.[87] Pensions have recently been held by the House of Lords to be the same type of receipt as private insurance benefits and therefore not deductible in assessing damages for loss of future earnings.[88]

(f) Interest on damages in personal injury cases

The discretionary power to award interest on damages granted to the courts by section 3 of the Law Reform (Miscellaneous Provisions) Act 1934, was made, in the absence of special reasons to the contrary, compulsory in respect of awards[89] for personal injuries by section 22 of the Administration of Justice Act 1969. That provision has now been repealed by the Administration of Justice Act 1982 which repeats the same provision with the addition that interest can still be awarded by the court even if the whole or part of the damages have already been paid.[90] In *Jefford* v. *Gee*,[91] the Court of Appeal established the principles to be applied in awarding interest. The general principle is that: "Interest should only be awarded to a plaintiff for being kept out of money which ought to have been paid to him."[92] The court decided therefore that on special damages interest should be awarded at half the appropriate

---

[85] Barnes v. *Bromley London Borough Council*, The Times, November 19, 1983; Denman v. *Essex Area Health Authority: Haste* v. *Sandell Perkins Ltd.* [1984] 3 W.L.R. 73; *The Times*, January 12, 1984.
[86] *Bradburn* v. *Great Western Ry.* (1874) L.R. 10 Ex. 1.
[87] *Hussain* v. *New Taplow Paper Mills Ltd.* [1987] I.C.R. 28; [1987] 1 All E.R. 417, C.A.
[88] *Parry* v. *Cleaver* [1970] A.C. 1, H.L.; reversing [1968] 1 Q.B. 195, C.A. Where the victim of an accident will have to remain in a National Health institution for life, it has been held that the benefit of having no living expenses is not to be taken into account as a deduction from damages for loss of earnings, as it is analogous either to private insurance or private benevolence. Public policy does not require the deduction of such benefits: *Daish* v. *Wauton* [1972] 2 Q.B. 262, C.A.
[89] Interest is only compulsory where there has been a "judgment." As to the position with regard to the various procedures that may follow where money is paid into court, see *Blundell* v. *Rimmer* [1971] 1 W.L.R. 123; [1971] 1 All E.R. 1072; *Newall* v. *Tunstall* [1971] 1 W.L.R. 105; [1970] 3 All E.R. 465, and *Waite* v. *Redpath Dorman Long Ltd.* [1971] 1 Q.B. 294.
[90] s.15 and Sched. 1.
[91] [1970] 2 Q.B. 130; [1970] 2 W.L.R. 702; [1970] 1 All E.R. 1202, C.A.
[92] *Ibid.*, per Lord Denning [1970] 2 W.L.R. 702 at 709; [1970] 1 All E.R. 1202 at 1208. The Court of Appeal applied this principle in deciding that interest should only be awarded against third parties joined to the action in respect of the period after they were joined: *Slater* v. *Hughes* [1971] 3 All E.R. 1287, C.A.

rate for the period from the date of the accident to the date of the trial.[93] With regard to general damages no award is appropriate in respect of damages for loss of future earnings.[94] Interest may, in appropriate cases, be awarded on the general damages for loss of amenity, pain, and suffering, even though an element in the assessment of those damages may have been an increase for the effect of inflation but, because an award of such damages takes account of the value of money at the date of trial, interest from the date of service of the writ to the date of trial must be at a very low rate, say 2 per cent.[95] Moreover, if there has been delay in prosecution of the action, interest will not be awarded up to the date of trial, but only up to the time that an action could have promptly been brought to trial.[96] Subject to the above rules, where an award of damages is increased on appeal, interest at the appropriate rate should be awarded on all the increase as from the date of the judgment at first instance, but this must take account of the 4 per cent. interest which automatically runs from this date on all judgment awards.[97]

(g) Exclusion clauses

It is theoretically possible for the right to sue for damages for personal injuries to be excluded or restricted (*e.g.* on a settlement) by agreement between the injured person and the tortfeasor(s). However, the Unfair Contract Terms Act 1977,[98] provides that it is not possible for any person carrying on a business (defined in s.14) by

---

[93] See *Dexter* v. *Courtaulds Ltd.* [1984] 1 All E.R. 70, C.A. Where total damages are limited by statute (*e.g.* by s.503 of the Merchant Shipping Act 1894 and s.3(2)(*a*) of the Merchant Shipping (Liability of Shipowners and Others) Act 1958), the whole of special damages will be taken to have been awarded, for interest purposes, and not just a rateable proportion—*McDermid* v. *Nash Dredging and Reclamation Co. Ltd.*, *The Times*, July 31, 1984.

[94] *Jefford* v. *Gee* was followed on this point in *Clarke* v. *Rotax Aircraft Equipment Ltd.* [1975] 3 All E.R. 794, C.A., and *Joyce* v. *Yeomans* [1981] 1 W.L.R. 549; [1981] 2 All E.R. 21, C.A. (no interest on damages for loss of earning capacity).

[95] *Wright* v. *British Railways Board* [1983] 2 All E.R. 698, H.L.; affirming *Birkett* v. *Hayes* [1982] 2 All E.R. 710, C.A. *Cf. Pickett* v. *British Rail Engineering Ltd.* [1980] A.C. 136, H.L., *Lim* v. *Camden and Islington Area Health Authority* [1979] 2 All E.R. 910, H.L.; and *Cookson* v. *Knowles* [1978] 2 All E.R. 604, H.L.

[96] *Birkett* v. *Hayes* [1982] 2 All E.R. 710, C.A. *Cf. McDermid* v. *Nash Dredging and Reclamation Co. Ltd.*, *The Times*, July 31, 1984 (affirmed on another point by C.A. [1986] Q.B. 965, and by H.L. [1987] 2 All E.R. 878; *The Times*, July 3, 1987).

[97] *Cook* v. *J.L. Keir & Co. Ltd.* [1970] 1 W.L.R. 774; [1970] 2 All E.R. 518, C.A.

[98] 1977 (c. 50), ss.1 and 2. See also the Employers' Liability (Defective Equipment) Act 1969, s.1(2).

any term in a contract[98a] made on or after February 1, 1978, or by any notice, to exclude or restrict liability for:
 (i) negligently[99] caused death or personal injury, or
 (ii) for negligently caused damage which is not death or personal injury, but in this latter case the exclusion or restriction is valid if proved to be reasonable.[1]

Moreover, a settlement by the employer's insurer with the employee is liable to be upset if the insurer broke the fiduciary duty of care owed to an employee acting without independent advice.[2]

(13) TIME LIMITS

(a) Limitation period

This is now governed by the Limitation Act 1980 which consolidates, and therefore repeals, all the previous Limitation Acts, so far as they affect the subject-matter of this work.

By the 1980 Act, an action for damages for personal injuries[3] must be brought within three years of either the date when the cause of action accrued or such later date as the plaintiff had "knowledge" of certain facts.[4] Thus an employee who cannot

---

[98a] Except:
 (a) any discharge and indemnity given by a person concerning an award to him of compensation for pneumoconiosis attributable to employment in the coal industry, in respect of any further claim arising from his contracting that disease—Sched. 1, para. 5; and
 (b) a contract of employment, except *in favour* of the employee—Sched. 1, para. 4.

[99] *i.e.* in breach of a duty in tort or contract to take reasonable care or exercise reasonable skill *or* the common duty of care imposed by the Occupiers' Liability Act 1957, s.1(1).

[1] The "reasonableness" test is defined in detail in s.11 of the Act.

[2] *Horry* v. *Tate and Lyle Refineries Ltd.* [1982] 2 Lloyd's Rep. 416; *The Times*, March 17, 1982.

[3] Includes "any disease and any impairment of a person's physical or mental condition, and 'injury' and cognate expressions shall be construed accordingly"—Limitation Act 1980, s.38(1). "Action" does not include an action previously statute-barred by the 12-months' period of limitation applicable to public authorities under s.21 of the Limitation Act 1939 and such an action was not 'revived' by the repeal of that section by the Limitation Act 1975 (introducing the rule now in s.11 of the 1980 Act)—*Arnold* v. *Central Electricity Generating Board* [1987] 3 All E.R. 694, H.L.

[4] Limitation Act 1980, s.11. The facts of which "knowledge," actual or constructive, is required before time starts to run are elaborately defined in s.14 of the 1980 Act and have been held to include the identity of the defendant—*Simpson*

reasonably realise, even with medical advice, that he is suffering from an industrial disease until many years after the first bodily harm was caused to him by the employer's negligence, etc., can sue within three years of acquiring "knowledge," without the need to ask for the leave of the court.[5] "In a case . . . where the acts and omissions on the part of the defendants which are complained of are, in broad terms, the exposure of their employee to dangerous working conditions and their failure to take reasonable and proper steps to protect him from such conditions, I think that the employee who has this broad knowledge may well have knowledge of the nature referred to [in the 1980 Act] sufficient to set time running against him, even though he may not yet have the knowledge sufficient to enable him or his legal advisers to draft a fully and comprehensively particularised statement of claim."[6] If he has died from the disease, the "knowledge" of his personal representative or dependants (under a Fatal Accident Acts claim) may then also become material.[7]

Even after three years have elapsed from both accrual of the cause of action and later knowledge, a plaintiff may now ask the court to override the limitation period altogether and allow the action to proceed.[8] The court in such a case has to have regard to a number of matters,[9] *e.g.* the reason for the delay, whether the defendant contributed to it, prejudice to the plaintiff or the

---

v. *Norwest Holst Southern* [1980] 2 All E.R. 471, C.A. Time does not run until the plaintiff has, *e.g.* knowledge that negligence, nuisance or breach of duty has caused injury that would reasonably justify proceedings (s.11). Belief is not the same as "knowledge"—*Davis* v. *Ministry of Defence, The Times*, August 7, 1985, C.A. (no "knowledge" by employee that dermatitis caused by work even though believed it to be so).

[5] *Ibid.*
[6] *Wilkinson* v. *Ancliff (BLT) Ltd.* [1986] 1 W.L.R. 1352 at 1365; [1986] 3 All E.R. 427 at 438, C.A.
[7] Limitation Act 1980, ss.13(1) and 11(5).
[8] Limitation Act 1980, s.33.
[9] Set out in s.33 of the 1980 Act. For a detailed consideration of such matters, see *Buck* v. *English Electric Co.* [1971] 1 W.L.R. 806; [1978] 1 All E.R. 271 (delay of 16 years no bar to widow's claim for husband's pneumoconiosis): *McCafferty* v. *Metropolitan Police District Receiver* [1977] 2 All E.R. 756, C.A. (delay because employee did not wish to jeopardise his job—excusable); *Firman* v. *Ellis* [1978] Q.B. 886; [1978] 2 All E.R. 851, C.A.; *Brooks* v. *J. & P. Coates Ltd.* [1984] I.C.R. 158; [1984] 1 All E.R. 702 and *Thompson* v. *Smiths Shiprepairers (North Shields) Ltd.* [1984] I.C.R. 236 (long delay no bar in actions for byssinosis and occupational deafness, respectively); *Thompson* v. *Brown & Co.* [1981] 1 W.L.R. 744; [1981] 2 All E.R. 296, H.L. (relevance of claim for negligence by plaintiff against his solicitors) and *Farmer* v. *National Coal Board, The Times*, April 27, 1985, C.A. (legal advice that claim would fail did not prevent subsequent action being time-barred).

defendant, and the nature of any medical or legal advice[10] received by the plaintiff.

For a plaintiff under a disability, *e.g.* minority or unsoundness of mind, at the date the cause of action accrued, the three-year period does not start to run until his disability ceases or he dies, whichever first occurs.[11]

If a defence of limitation is relied upon it must be pleaded and applications for late amendment of a defence for this purpose may well be refused.[12]

(b) Dismissal for want of prosecution

Even where the action has been commenced in time, the court may thereafter strike the action out if there is delay in prosecuting the action to trial.[13] The authorities were reviewed by the House of Lords in *Birkett* v. *James*,[14] where it was said that an action should not be dismissed (for want of prosecution) before the limitation period had expired, save in rare and exceptional circumstances because the plaintiff may merely start a second action, if his first action is dismissed within the limitation period.[15] It was held that, where the application to dismiss is after the expiry of the limitation period, an action can be dismissed only if the delay after the issue of the writ:

(a) exceeds the time-limits prescribed by the rules of court,
(b) is inordinate and inexcusable having regard to the delay before the issue of the writ and
(c) has increased, by more than a minimal amount, the prejudice already suffered by the defendant by reason of the delay in issuing the writ.

---

[10] The plaintiff can be required to divulge the nature of the legal advice, despite legal professional privilege—*Jones* v. *G.D. Searle Co. Ltd.* [1978] 3 All E.R. 851, C.A.
[11] *Ketteman* v. *Hansel Properties Ltd.*, *The Times*, January 23, 1987, H.L.
[12] *Ibid.*
[13] See *Allen* v. *Sir Alfred McAlpine and Sons Ltd.* [1968] 2 Q.B. 229; [1968] 1 All E.R. 543, C.A. For the extent of a solicitor's liability to his client in such circumstances, see *Mainz* v. *Dodd*, *The Times*, July 21, 1978.
[14] [1977] 2 All E.R. 801, H.L.
[15] Thus an action by a minor (infant) brought during minority should not be struck out because the minor has a right to bring the action within the requisite period after attaining full age (*e.g.* three years after 18th birthday for personal injuries action)—*Tolley* v. *Morris* [1979] 2 All E.R. 561, H.L. However, a second action may also be struck out as an abuse of the process of the court, even though the second action is brought within the limitation period—*Janor* v. *Morris* [1981] 3 All E.R. 780, C.A.

Birkett v. *James* was distinguished by the Court of Appeal in *Biss* v. *Lambeth, Southwark and Lewisham Area Health Authority (Teaching)*[16] as being concerned only with cases where there has been no extension of time for issue of the writ in a personal injuries action, under the Limitation Act 1980. Where such an extension of time has been granted, delay after the issue of the writ, even if the time-limits in the rules of court are not exceeded, may cause the action to be struck out for want of prosecution.

(14) COMPULSORY INSURANCE

Under the Employers' Liability (Compulsory Insurance) Act 1969, which came into force on January 1, 1972, every employer carrying on business in Great Britain must insure against liability for personal injury arising out of and in the course of employment in Great Britain in his business.[17]

Certain specified large employers are excluded from this obligation by the Act and regulations made under the Act.[18] Regulations have also detailed the nature of approved policies and authorised insurers for the purposes of the Act. Failure to hold the insurance required by the Act is a criminal offence. Display of copies of certificates of insurance is required at each place of business where employees whose claims may be subject to an indemnity under the insurance policy are employed. Health and Safety Executive officers may require employers to produce copies of such certificates.

The insurer owes to an employee *acting without independent advice* a fiduciary duty of care in settling the employee's claim. Such a settlement may therefore be upset and is not a defence to a claim by the employee against the employer for a larger sum.[19]

---

[16] [1978] 1 W.L.R. 382; [1978] 2 All E.R. 125, C.A. (nine months' delay after writ issued, with leave of court, 10 years after alleged negligence causing personal injuries—action struck out for want of prosecution).

[17] For criticism of gaps in the protection for employees (*e.g.* voidability of the policy and its being subject to conditions), see *Dunbar* v. *A. & B. Painters Ltd.*, *The Times*, March 14, 1986, C.A.

[18] See the Employers' Liability (Compulsory Insurance) Exemption Regulations 1971 (S.I. 1971 No. 1933), as amended by S.I.s 1974 No. 208, 1975 No. 194 and 1981 No. 1489. As to offshore installations, see S.I. 1975 Nos. 1289 and 1443.

[19] *Horry* v. *Tate and Lyle Refineries Ltd.* [1982] 2 Lloyd's Rep. 416; *The Times*, March 17, 1982.

(15) CLAIMS BY EMPLOYEES FOR SALARY, WAGES, STATUTORY SICK PAY, ETC.

(a) Introductory

If an employee is incapacitated by industrial disease or accident, *i.e.* a disease which has developed as a result of working conditions or an accident sustained at work, there are two possible types of claim that he can make. One, dealt with here, is the claim against the employer for salary, wages, statutory sick pay, etc. These claims can be maintained either under the contract of employment or under the relevant Social Security legislation. Such claims can be maintained irrespective of whether or not the accident or disease was caused by negligence or breach of statutory duty by the employer. If negligence or breach of statutory duty can be shown, then of course the employee may in addition be able to claim damages in tort and those damages may include loss of wages both past and future. But damages for the same item would not be awarded twice by the courts. Another situation that may arise is that an employee is suspended on medical grounds under the provisions of various Health and Safety Regulations and there is then a limited right for the employee to claim his wages, salary, etc., from the employer.[20]

(b) Claims for salary and wages while incapacitated

If an employee is away from work while incapacitated he may be able to claim salary or wages from his employer. If as well as being caused by the work, the injury or the disease was the result of negligence or breach of statutory duty by the employer or those for whom he is responsible[21] then the claim for salary or wages can be maintained in tort irrespective of whether there is any right (implied or expressed) in the contract of employment for the payment in such circumstances. If however there is no such negligence or breach of statutory duty, the claim will have to be made for payment under an express or implied term of the contract of employment. It is irrelevant, in ascertaining whether there is such an express or implied term, that the incapacity for work arose out of working conditions. There is no right to a contractual payment just because the injury or disease arose out of working conditions. If there is an express term for payment of salary or wages during

---

[20] See pp. 58–60.
[21] See Chap. 2, ss.(1) to(14) and Chap. 3.

sickness then of course no problem arises save as to the construction of the express term. Commonly contracts of employment for the payment of full salary or wages for, *e.g.* the first six months of incapacity, then half salary or wages for say a further six months and thereafter nothing. If there is no express term, then it has to be ascertained whether there is an implied term of the contract for payment. This depends on whether or not truly the remuneration under the contract is paid for the claimant's being available for work when fit to do so or only for actual work done. There is no presumption that wages are payable during sickness.[22]

(c) Statutory sick pay

Under the Social Security and Housing Benefits Act 1982 a new system was introduced by which employers became liable to pay to their sick employees "statutory sick pay" whatever the cause of incapacity for work.[23] Any agreement in the contract of employment to exclude that right or vary it or require the employee to contribute towards any costs incurred by the employer is void.[24] Moreover it is provided[25] that "statutory sick pay may not be paid in kind or by way of provision of board or lodging or of services or other facilities." However, statutory sick pay is not necessarily as full in time of payment or amount as wages under the contract. There is a maximum period in any one tax year (April 5 in one year to April 6 in next) of twenty-six weeks' statutory sick pay.[26–27] Statutory sick pay is not payable for the first three qualifying days in any period of entitlement ("waiting days"). Moreover, the amount of statutory sick pay is not the full amount of wages but is a lesser figure, the current rates being as follows:[28]

Wages	Statutory Sick Pay
£76.50 per week or more	£47.20 per week
£39 to £76.49 per week	£32.85 per week

Once the employer has paid statutory sick pay to his employee he is able to recoup the amount he has paid by deduction from the

---

[22] *Mears* v. *Safecar Security* [1983] Q.B. 54, C.A.
[23] Provided of course that the incapacity is "by reason of some specific disease or bodily or mental disablement for doing work which [the employee] can reasonably be expected to do under [the contract of employment]"—1982 Act, s.1(3).
[24] 1982 Act, s.1(2).
[25] By reg. 8 of the Statutory Sick Pay (General) Regulations 1982 (S.I. 1982 No. 894).
[26–27] 1982 Act, s.5(4) as amended.
[28] 1982 Act, s.7(1) as amended.

employer's share of social security contributions paid on behalf of the employee.[29] Disputes as to questions of entitlement to statutory sick pay are determined by the social security adjudicating authorities, *i.e.* local adjudication officers, Social Security Appeal Tribunals, Social Security Commissioners and thence appeal to the Court of Appeal and House of Lords in the same manner as other social security questions and claims.[30] This represents a unique departure in that the social security adjudicating authorities are deciding disputes between employer and employee, such disputes having previously been decided either by the courts or by industrial tribunals.[31]

(d) Suspension from work on medical grounds

An employee who is not himself incapable of work by reason of illness, etc., may nevertheless have to be suspended by an employer from work for precautionary medical reasons as a result either of an Act of Parliament dealing with health and safety at work, *e.g.* the Health and Safety at Work, etc., Act 1974 or as a result of a Health and Safety Regulation or as a result of a recommendation in a Code of Practice issued by the Health and Safety Commission.[32] A statutory right of the employee to be paid remuneration during certain kinds of suspension is to be found in sections 19 to 22 of the Employment Protection (Consolidation) Act 1978 which provides for the payment of such remuneration for a period not exceeding 26 weeks.[33] The suspensive provisions are those which are set out in Schedule 1 to the 1978 Act[34] and are as follows:

> "The India Rubber Regulations 1922 S.R. and O. 1922 No. 329 regulation 12.
> The Chemical Works Regulations 1922 S.R. and O. 1922 No. 731 regulation 30.
> The Ionising Radiations (Unsealed Radioactive Substances) Regulations 1968 S.I. 1968 No. 780 regulations 12 and 33.
> The Ionising Radiations (Sealed Sources) Regulations 1969 S.I. 1969 No. 808 regulations 11 and 30.

---

[29] 1982 Act, s.9.
[30] See pp. 107–109.
[31] Compare Chap. 6 (Industrial Tribunals).
[32] Under s.16 of the Health and Safety at Work, etc., Act 1974.
[33] 1978 Act, s.19(1).
[34] As amended by the Employment Protection (Medical Supervision) Order 1980 (S.I. 1980 No. 1581).

The Radioactive Substances (Road Transport Workers) (Great Britain) Regulations 1970 (as amended by S.I. 1975 No. 1522) regulation 14.
The Control of Lead at Work Regulations 1980 S.I. 1980 No. 1248 regulation 16."

Codes of Practice made under these Regulations are also included. An employee cannot claim remuneration during suspension under any of the above provisions unless he has been continuously employed for a period of not less than one month ending with the day before that on which the suspension begins.[35] The employee must still be capable of work.[36] The employee will be regarded as suspended only so long as he is under contract of employment with his employer. If the employee is not provided with work or does not perform the work he normally performed before the suspension he may make a claim for remuneration, though he does of course have to give credit for any remuneration actually received from the employer.[37] Moreover an employee is not entitled to make a claim under this head if he unreasonably refuses to perform alternative work offered to him by his employer which is suitable for him or alternatively does not comply with reasonable requirements imposed by his employer with a view to ensuring that his services are available, *e.g.* presumably not going abroad on holiday.[38]

If instead of suspending the employee on medical grounds, the employer dismisses an employee because of a requirement or recommendation in a Regulation or Code of Practice to suspend on medical grounds then the qualifying period for the employee to claim unfair dismissal in an industrial tribunal instead of being the usual two years[39] is reduced to one month's continuous employment.[40]

Disputes as to entitlement to remuneration under this provision or as to its amount are to be taken to an industrial tribunal for

---

[35] 1978 Act, s.20(1) as inserted by the Employment Act 1982—this period may be extended to three months if it were a contract for a fixed term of three months or less or a contract to do a specific task not expected to last for more than three months—1978 Act, s.20(2) as inserted by 1982 Act.
[36] 1978 Act, s.20(3)—if he is incapable of work then his claims against the employer must lie under some other head—see, above, pp. 56–57.
[37] 1978 Act, s.21.
[38] 1978 Act, s.20(4).
[39] 1978 Act, s.64(1)(*a*).
[40] 1978 Act, s.64(2).

determination and are not for the ordinary courts.[41] An employee may of course claim in the ordinary courts for wages under the contract if an express or implied term of the contract provides to that effect.[42]

---

[41] 1978 Act, s.22. The claim to an industrial tribunal must be made within three months of the day for which remuneration is claimed or such further period as is considered reasonable by a tribunal when satisfied it was not reasonably practicable for the complaint to be brought within the three months' period.

[42] See pp. 56–57.

3. Liability of Employer for Damages for Breach of Statutory Duty

(1) INTRODUCTORY

In addition to his duty at common law, an employer has to observe many provisions relating to the safety, health and welfare of his employees which are laid down by statute, the breach of which may make him liable to prosecution as well as to an action for damages for any injury caused by the breach. Where an employee is injured in circumstances which constitute both negligence at common law and a breach of statutory duty, he can and usually will put forward a claim under both heads, for they are in no sense exclusive of each other.[1] The question whether or not a given set of facts constitutes a breach of statutory duty is one which must always depend on the particular wording of the Act or regulation concerned. The principal areas of employment covered by such legislation are factories (see the Factories Act 1961 and regulations thereunder); offices, shops and railway premises (see the Offices, Shops and Railway Premises Act 1963 and regulations thereunder); mines and quarries (see the Mines and Quarries Act 1954 and regulations thereunder); agriculture (see the Agriculture (Safety, Health and Welfare Provisions) Act 1956 and regulations thereunder); merchant shipping and fishing vessels[2]; and offshore

---

[1] In *Bux* v. *Slough Metals Ltd.* [1973] 1 W.L.R. 1358; [1974] 1 All E.R. 262, C.A., the Court of Appeal rejected a submission that once a finding of no breach of statutory duty was made, the defendant could not be liable at common law. Reference was made to a solid body of authority to the contrary. *Per* Stephenson L.J.: "There is . . . no presumption that a statutory obligation abrogates or supersedes an employer's common law duty, or that it defines or measures his common law duty either by clarifying it or cutting it down—or indeed by extending it. It is not necessarily exhaustive of that duty or co-extensive with it . . . The statutory obligation may exceed the duty at common law or it may fall short of it, or it may equal it."

[2] See the Merchant Shipping Act 1970 (c. 36), ss.19–26; the Merchant Shipping Act 1979 (c. 39), ss.21–23 and the Safety at Sea Act 1986 (c. 23).

installations, *e.g.* oil-rigs (see the Mineral Workings (Offshore Installations) Act 1971, the Petroleum and Submarine Pipe-lines Act 1975 and the Oil and Gas (Enterprise) Act 1982 and regulations thereunder). Breach of duties under all this legislation can result in criminal prosecution and, in most cases, civil liability for damages. In addition, the duties in sections 1–9 of the Health and Safety at Work etc. Act 1974, are general in scope and are not confined to particular areas of employment. However, breach of them, though a criminal offence cannot *per se* give rise to a civil action for damages. It should also be remembered that statutes in areas other than those of employee protection may impose duties on employers, which could have some relevance in an action brought by an employee or other person injured by the employer's breach of duty. Examples of such statutory provisions are the Nuclear Installations Act 1965 and regulations made thereunder, the Gas Act 1972 and safety regulations made under that Act,[3] and the Reservoirs (Safety Provisions) Act 1930 and Reservoirs Act 1975.

Where an employer is liable to his employee for damages for breach of statutory duty, he may have a right of contribution or indemnity against some other person.[4]

What is proposed in this chapter is to give a general account of the working of the statutory safety code, as well as to refer to those rules and phrases which are either of common or recurrent application throughout the various statutory provisions.

In doing this, it will be convenient to divide the subject-matter into the four headings which ordinarily arise for consideration when a claim for damages for breach of statutory duty arises: that is:

   (i) the nature of the duty owed;
   (ii) the standard required;
   (iii) proof of the claim; and
   (iv) defences to the claim.

(a) Who can sue

Unlike common law negligence, the duty owed by an employer is not necessarily confined to his own employees nor does it necessarily extend to all his employees. The question whether or not an injured person has a right of action for breach of statutory duty

---

[3] s.47(1). Breach of duties under regulations made under the 1974 Act will give rise to civil liability for damages—s.47(2).
[4] See pp. 8–9 above.

must in all cases depend on the precise wording of the Act or Regulations concerned, and this frequently is a matter of some difficulty.[5]

A common statutory description of the class of person intended to be protected is "person employed." This phrase includes the employee of an independent contractor who is working on the premises or a factory occupier.[6] It has also been held to include anyone who is working in a factory, no matter by whom he is employed.[7] But the position of an independent contractor working in person is unsettled; under the Building Regulations, 1948 (where the duty is defined in somewhat different terms), it has been held that he cannot recover against a main contractor for a breach of the regulations.[8] A fireman coming to put out a fire at a factory is not "a person employed," so far as the statutory duty of the factory occupier is concerned,[9] nor is an employee who is doing private work at a factory in his own time.[10] But the fact that an employee is in a part of office premises other than that in which he is employed to work does not take him outside the protection of the Offices, Shops and Railway Premises Act. Even if he has been forbidden to go there he retains this protection because the statutory duties are not confined to protecting those injured in the course of their employment.[11]

Where the qualifying words are "person employed in the process," a more limited view obtains as to the category of persons protected. The words have been held to apply only to persons employed by the occupier of the premises in the processes of the occupier.[12] An employee may still be "employed in" a particular type of work, even though it was not part of his ordinary duties and he had volunteered to do it on the particular occasion, if in the past his doing it voluntarily had been accepted by the employer.[13]

---

[5] As to prisoners injured, while training in prison to use machinery, see *Ferguson v. Home Office*, The Times, October 8, 1977 (Home Office liable for inadequate training in use of machine saw).
[6] *Stanton Iron Works v. Skipper* [1956] 1 Q.B. 255; [1955] 3 All E.R. 544, as explained in *Canadian Pacific v. Bryers* [1958] A.C. 485; [1957] 3 All E.R. 572, H.L.
[7] *Massey-Harris-Ferguson v. Piper* [1956] 2 Q.B. 396; [1956] 2 All E.R. 722; approved by the House of Lords in *Canadian Pacific v. Bryers*, *supra*.
[8] *Herbert v. Harold Shaw* [1959] 2 Q.B. 138; [1959] 2 All E.R. 189.
[9] *Hartley v. Mayoh* [1954] 1 Q.B. 383; [1954] 1 All E.R. 375, C.A. though if the fire has been negligently caused by some person, that person can be sued by an injured fireman attending the fire—*Salmon v. Seafarer Restaurant* [1983] 1 W.L.R. 1264; [1983] 3 All E.R. 729.
[10] *Napieralski v. Curtis (Contractors)* [1959] 1 W.L.R. 835; [1959] 2 All E.R. 426.
[11] *Westwood v. Post Office* [1974] A.C. 1, H.L.
[12] *Whalley v. Briggs Motors* [1954] 1 W.L.R. 840; [1954] 2 All E.R. 193.
[13] *Thompson v. National Coal Board* [1982] I.C.R. 15, C.A.

Another form of qualifying words is "where any person has at any time to work." This has been held to include both an independent contractor in person[14] and the employee of an independent contractor,[15] when working on the premises of a factory occupier.

Difficult questions may arise when regulations have been made under one Act, and the class of persons for whose benefit regulations may be made is subsequently enlarged by a further statute. Further difficulties may arise when there appears to be some conflict between the class of persons envisaged by a regulation, and the class authorised by the enabling statute. In this connection, the following general rules have been laid down by the House of Lords[16]:

1. Regulations cannot be made in favour of persons outside the class authorised by statute.
2. Regulations may define and narrow the class authorised by statute.
3. Regulations made under one Act cannot receive a wider construction at some later date merely because another Act has been passed altering the class of persons for whose benefit the regulations may be made.

(b) Who can be sued

Where the statutory liability is laid on the occupier of the premises, there may sometimes be different occupiers for the purposes of different statutory provisions, in which case both will be liable for a breach of statutory duty. If the duty is expressed to rest on the occupier, he will be liable for a breach of duty, even though that breach is solely caused by some third party, such as an independent contractor[17] or an electricity company supplying power to a factory.[18]

A seller of dangerous machinery cannot be made liable in a claim for personal injuries by an injured employee for breach of the statutory duty (under s.17(2) of the Factories Act 1961) to supply encased gearing, etc.,[19] but he will be liable to the

---

[14] *Lavender* v. *Diamints* [1949] 1 K.B. 585; [1949] 1 All E.R. 532, C.A.
[15] *Whitby* v. *Burt Boulton* [1947] K.B. 918; [1947] 2 All E.R. 324; *Whincup* v. *Woodhead (Joseph) & Sons (Engineers)* [1951] 1 All E.R. 387; 115 J.P. 97.
[16] *Canadian Pacific* v. *Bryers* [1958] A.C. 485; [1957] 3 All E.R 572, H.L.
[17] *Whitby* v. *Burt Boulton* [1947] K.B. 918; [1947] 2 All E.R. 324; *Hosking* v. *De Havilland Aircraft Co.* [1949] 1 All E.R. 540; 83 Ll.L.Rep. 11; *Braham* v. *J. Lyons & Co.* [1962] 1 W.L.R. 1048, at 1051; [1962] 3 All E.R. 281, C.A., at 283, *per* Lord Denning, M.R.
[18] *Heard* v. *Brymbo Steel Co.* [1947] K.B. 692; 80 Ll.L.Rep. 424, C.A.
[19] *Biddle* v. *Truvox* [1952] 1 K.B. 101; [1951] 2 All E.R. 835.

employee for damages in the tort of negligence if the machinery was faultily manufactured.[20]

Where the employer is the Crown,[21] it is basically under the same liability to its employees as any other employer,[22] including vicarious liability for its employees' torts,[23] though the Crown is not bound by a statutory duty unless the statute expressly, or by necessary implication, so enacts. The Crown was normally not liable in tort for injury sustained by members of the armed forces, though a civilian hospital to which an injured serviceman was taken might be[24] but the Crown Proceedings (Armed Forces) Act 1987 has abolished that rule for the future.[25] Commonly, statutes in this field bind the Crown but enact special rules in relation to it.[26] Foreign states are liable to be sued by their employees for breaches of statutory duty causing death or personal injury, provided the breaches consisted of acts or omissions in the United Kingdom.[27]

(c) Jurisdiction

It has been held that the Docks Regulations, 1934, do not apply to places outside the jurisdiction,[28] and in the absence of some special provision a similar result would apply in the case of other statutory safety provisions. However, the Health and Safety at Work, etc., Act 1974[29] (other than Part III, which is concerned

---

[20] See *e.g. Hill* v. *James Crowe (Cases) Ltd.* [1978] 1 All E.R. 812.
[21] *i.e.* not the Sovereign personally, but includes, *e.g.* Government Departments, H.M. Forces and a number of other bodies such as the Forestry Commission, Medical Research Council, etc. Nationalised industries are not Crown bodies neither, for the purposes of health and safety legislation, is the National Health Service (National Health Service (Amendment) Act 1986).
[22] Crown Proceedings Act 1947 (c. 44), s.2(2).
[23] *Ibid.* s.2(3).
[24] See Crown Proceedings Act 1947 (c. 44), s.10, *Bell* v. *Secretary of State for Defence, The Times* June 29, 1985, C.A. and *Pearce* v. *Secretary of State for Defence, The Times,* December 31, 1986.
[25] Crown Proceedings (Armed Forces) Act 1987 (c. 25), s.1, repealing s.10 of the Crown Proceedings Act 1947 (see previous footnote) but not "in relation to anything suffered by a person in consequence of an act or omission committed before [May 15, 1987]." A power is given by s.2 of the 1987 Act to the Secretary of State to revive s.10 of the 1947 Act in certain circumstances.
[26] See, *e.g.* the Factories Act 1961, s.173, the Health and Safety at Work etc. Act 1974, s.48; the Agriculture (Safety, Health and Welfare Provisions) Act 1956, s.22, and the Offices, Shops and Railways Premises Act 1963, s.83.
[27] State Immunity Act 1978 (c. 33), ss.4 and 5. For the definition of "State," see s.14.
[28] *Yorke* v. *British and Continental S.S. Co.* (1945) 78 Ll.L.Rep. 181, C.A.
[29] 1974, c. 37.

with building regulations) applies, by virtue of the Health and Safety at Work etc. Act 1974 (Application outside Great Britain) Order 1977[30] to certain offshore installations, etc., and to certain activities in territorial waters.

(d) Rules of construction

Since a breach of statutory duty generally constitutes a criminal offence as well as a civil wrong, it has been said that, in cases of doubtful construction, that construction is to be preferred which avoids the creation of an offence.[31] But it has also been pointed out that one of the objects of the Factories Acts and the regulations passed thereunder is to protect workmen, and that they must therefore be read so as to effect the object so far as the wording fairly and reasonably permits[32]; and that the Acts should be construed so as to further the end of preventing accidents to workmen.[33] It has also been said that the rule of construction against creating offences is a rule of last resort, and only to be applied when other rules fail.[34]

If a construction of a phrase is put forward which, if correct, would make the whole process incapable of being carried out, that construction must be rejected, if the words admit of any doubt. But where the meaning of the words is plain, the fact that compliance with the statutory provision would make the plant concerned commercially unusable cannot affect the duty imposed; thus the fact that, in order to attain secure fencing within the fencing provisions of the Factories Acts, a grindstone would have to be so fenced as to be commercially unusable was held not to affect the obligation to fence, which was absolute.[35]

If a word or phrase has been given a certain meaning in a decision of the courts, and subsequently thereto the word or phrase concerned has been reproduced in an Act of Parliament dealing with the same subject-matter, the legislature will be presumed to have endorsed the meaning so given.[36]

---

[30] S.I. 1977 No. 1232.
[31] *Burns* v. *Terry* [1951] 1 K.B. 454; [1950] 2 All E.R. 987, C.A.; *Rees* v. *Bernard Hastie* [1953] 1 Q.B. 328; [1953] 1 All E.R. 375, C.A.; *Sumner* v. *Priestley* [1955] 1 W.L.R. 1202; [1955] 3 All E.R. 445, C.A.
[32] *Harrison* v. *N.C.B.* [1951] A.C. 639; [1951] 1 All E.R. 1102, H.L., per Lord Porter.
[33] *Norris* v. *Syndic* [1952] 2 Q.B. 135; [1952] 1 All E.R. 935, C.A.
[34] *McCarthy* v. *Coldair* [1951] W.N. 590; [1951] 2 T.L.R. 1226, C.A.
[35] *Summers* v. *Frost* [1955] A.C. 740; [1955] 1 All E.R. 870, H.L.; *Mackay* v. *Culsa Shipbuilding Co.*, 1945 S.C. 414; 1946 S.L.T. 104.
[36] *Hamilton* v. *N.C.B.* [1960] A.C. 633; [1960] 1 All E.R. 76, H.L.

(e) Limitation of duty to particular class of injury

In one case dust from monsonia wood caused the plaintiff to contract dermatitis, and his claim under section 4 of the Factories Act, 1937 (now s.4 of the Factories Act 1961), which dealt with protection against dust failed, as the section was only aimed at preventing injuries arising through inadequate ventilation.[37]

The general test in this connection is that propounded by Lord Reid; if the words of the statute indicate that they are only aimed at preventing a certain kind of injury, then civil liability will reasonably be confined to such injury.[38] In one case, injury was caused by wire ejected from a machine. There was no statutory duty to fence against this ejected wire, but the machine was in fact unguarded altogether, and if a guard had been provided in compliance with the duty to fence the machine so as to keep people out, it would also incidentally have kept the wire in. It was held that no claim lay for breach of statutory duty, for although the defendants were in breach of their duty to fence so as to keep people out, the injury in question was not of the kind aimed at by the statute.[39] A similar type of question also arose in another case[40] in connection with foundry dust, of which there were two types, the visible but non-injurious dust, and the invisible injurious dust. There was no duty to give protection against the latter, because at the time it was unknown, but if the defendants in pursuance of their duty to protect against the former had provided a certain type of respirator, this would, albeit incidentally, have given protection against the latter. The defendants were held liable. In a Scottish case, the court rejected an argument that the duty in s.29(1) of the Factories Act 1961 to make and keep a safe place of work was not applicable to a high noise level emanating from a machine.[41] In an earlier English case the point had been left open.[42]

---

[37] *Ebbs* v. *James Whitson* [1952] 2 Q.B. 877; [1952] 2 All E.R. 192, C.A.; followed in *Graham* v. *Co-operative Wholesale Society* [1957] 1 W.L.R. 511; [1957] 1 All E.R. 654, and impliedly approved by the House of Lords in *Nicholson* v. *Atlas Steel Foundry and Engineering Co.* [1957] 1 W.L.R. 613; [1957] 1 All E.R. 776, H.L.; 1957 S.C., H.L. 44.
[38] *Grant* v. *N.C.B.* [1956] A.C. 649; [1956] 1 All E.R. 682, H.L.; 1956 S.C., H.L. 48.
[39] *Kilgollan* v. *W. Cooke* [1956] 1 W.L.R. 527; [1956] 2 All E.R. 294, C.A.; applying *Grant* v. *N.C.B.*, *supra*. See also *Carroll* v. *Andrew Barclay & Sons*, 1948 S.C.,H.L. 100; [1948] A.C. 477.
[40] *Richards* v. *Highway Ironfounders* [1957] 1 W.L.R. 781; [1957] 2 All E.R. 162; cf. *Clifford* v.*Challen* [1951] 1 K.B. 495.
[41] *Carragher* v. *Singer Manufacturing Co. Ltd.*, 1974 S.L.T. (Notes) 28.
[42] *Berry* v. *Stone Manganese Marine Ltd.* [1972] 1 Lloyd's Rep. 182.

(2) STANDARD OF CARE REQUIRED

(a) "Dangerous"

Frequently a statutory provision requires protection against "danger" or a "dangerous part." Many, if not most, of these cases arise under the fencing provisions of the Factories Acts, which should be referred to, but there are certain general rules which are commonly applied by the courts in considering "danger" and "dangerous."

Thus, "in considering whether machinery is dangerous, it must not be assumed that everyone will always be careful. A part of a machine is dangerous if it is a possible cause of injury to anybody acting in a way in which a human being may reasonably be expected to act in circumstances which may reasonably be expected to occur."[43] But, as has been pointed out by the Divisional Court,[44] in applying this dictum, "nobody expects human beings to be so extravagantly careless as to touch some part of the machine which is not only not difficult to avoid, but which is actually difficult to get near."[45] "Danger" does not mean exceptional danger, but refers to all ordinary danger.[46] Thus a "repeat" or an uncovenanted stroke by a machine will probably fall into the former category, so that the employer will not be liable, unless there is proof that the machine had a propensity to repeat.[47]

Experienced men must be protected from danger as well as inexperienced, for both are equally liable to lapses.[48]

(b) "Safe"

"Safe" means safe for all contingencies that may reasonably be foreseen, unlikely as well as likely, possible as well as probable.[49]

---

[43] *Walker* v. *Bletchley Flettons* [1937] 1 All E.R. 170, *per* du Parcq J., approved by the House of Lords in *Summers* v. *Frost* [1955] A.C. 740; [1955] 1 All E.R. 870, H.L.
[44] *Carr* v. *Mercantile Produce* [1949] 2 K.B. 601; [1949] 2 All E.R. 531.
[45] *Walker* v. *Bletchley Flettons*, also *per* du Parcq J., *supra*.
[46] *Hutchinson* v. *L.N.E.R.* [1942] 1 K.B. 481; [1941] 1 All E.R. 330, C.A.
[47] *Eaves* v. *Morris Motors* [1961] 3 W.L.R. 657; [1961] 3 All E.R. 233, C.A.
[48] *Murfin* v. *United Steel Companies* [1957] 1 W.L.R.104; [1957] 1 All E.R. 23, C.A.
[49] *McCarthy* v. *Coldair* [1951] W.N. 590; [1951] 2 T.L.R. 1226, C.A.

To be safe a means of access must be free from any danger which is both appreciable and foreseeable.[50]

(c) "Secure" and "securely"

A duty to fence securely, or to secure a place of work against danger, unless the obligation is expressly qualified, is an absolute duty, and it is no defence to say that compliance with the obligation would make the object in question commercially unusable.[51] Following this, it has been held that security means absolute protection against all ordinary hazards, subject only to the test of foreseeability.

Where there is an obligation to fence something securely:

> " . . . the standard to be observed is a fence which will prevent accidents occurring in circumstances which may reasonably be anticipated, but the circumstances which may be anticipated reasonably include a great deal more than the staid, prudent, well-regulated conduct of men diligently attentive to their work, and the occupiers of factories are bound to reckon on the possibility of conduct very different from that. They are bound to take into account the possibility of negligent, ill-advised or indolent conduct on the part of their employees, and even of frivolous conduct, especially where young persons are employed."[52]

Accordingly, the test of foreseeability is to some extent a criterion in determining whether a thing is or is not secure.[53] Thus a good and solid ladder does not cease to be a secure foothold merely because the plaintiff in using it needed to use both his hands in cleaning a window. The true test was to ask whether the ladder afforded a secure foothold to an experienced window cleaner acting in a way that such a man might reasonably be expected to act in circumstances that might reasonably be expected to occur, and applying this test the ladder was held to be secure.[54] A place or object is not "secure" if it contains a latent defect of a type which is intrinsically likely to occur in the course of the work concerned.[55]

---

[50] *Moodie* v. *Furness Shipbuilding* [1951] 2 Lloyd's Rep. 600, C.A.
[51] *Summers* v. *Frost* [1955] A.C. 740; [1955] 1 All E.R. 870, H.L.
[52] *Smith* v. *Chesterfield Co-operative Society* [1953] 1 W.L.R. 370; [1953] 1 All E.R. 447, C.A.
[53] *Burns* v. *Terry* [1951] 1 K.B. 454; [1950] 2 All E.R. 987, C.A., approved by the House of Lords in *Summers* v. *Frost* [1955] A.C. 740; [1955] 1 All E.R. 870.
[54] *Wigley* v. *British Vinegars* [1962] 3 W.L.R. 731; [1962] 3 All E.R. 161, H.L.
[55] *Marshall* v. *Gotham* [1954] A.C. 360.

(d) "Practicable" and "reasonably practicable"

It is common to find a statutory obligation which requires the employer to take all "practicable" measures, or to take precautions "so far as is reasonably practicable."[56]

It would appear that it is not "practicable" to take precautions against a danger which at the material time is wholly unknown.[57] If a defendant is under a duty to take "all practicable measures" to obviate a risk, this may well include a duty to supervise and enforce the use of protective equipment provided.[58]

"Impracticable," on the other hand, is occasionally found, and is used to indicate relief from statutory obligation. The word has been interpreted strictly, and held to mean more than "not reasonably practicable."[59] The fact that compliance with the obligation would involve unreasonable time or expense is irrelevant.[60] But there is a greater difference between "impracticable" and "impossible." The former at all events introduces some degree of reason and involves some regard for practice.[61]

"Reasonably practicable" imposes a less stringent standard than "practicable." There may well be precautions which it is practicable but not reasonably practicable to take.

In considering what is reasonably practicable, the employer must have regard to the period of time over which the danger is spread, and the time, trouble and expense of the safeguards which would be required; if the latter are disproportionate to the period of the risk, it will not be reasonably practicable to take them.[62] "Reasonably practicable" is narrower than "physically possible," and implies a computation between quantum of risk on the one hand and time, trouble and expense of safeguards on the other, and if the defendant can show a gross disproportion (*i.e.* a very small risk against inordinate safeguards), he discharges his duty.[63]

---

[56] See, for example, ss.28(1) and 29(1) of the Factories Act, 1961 (safe floors, stairs, gangways etc. *and* safe means of access, respectively).
[57] *Adsett* v. *K. & L. Steelfounders* [1953] 1 W.L.R. 773; [1953] 2 All E.R. 320, C.A.; *Richards* v. *Highway Ironfounders* [1955] 1 W.L.R. 1049; [1955] 3 All E.R. 205, C.A.; *Gregson* v. *Hick Hargreaves & Co.* [1955] 1 W.L.R. 1252; [1955] 3 All E.R. 507, C.A.
[58] *Crookall* v. *Vickers-Armstrong* [1955] 1 W.L.R. 659; [1955] 2 All E.R. 12.
[59] *Brown* v. *National Coal Board* [1962] A.C. 574, H.L., at p. 598; [1961] 1 Q.B. 303, C.A., at 332; *Jayne* v. *National Coal Board* [1963] 2 All E.R. 220, at 223.
[60] *Moorcroft* v. *Thomas Powles & Sons, Ltd.* [1962] 3 All E.R. 741, D.C.; *Cork* v. *Kirby Maclean, Ltd.* [1952] 1 All E.R. 1064.
[61] *Jayne* v. *National Coal Board* [1963] 2 All E.R. 220, at 223.
[62] *Coltness Iron Co.* v. *Sharp* [1938] A.C. 90; [1937] 3 All E.R. 593, H.L.
[63] *Edwards* v. *N.C.B.* [1949] 1 K.B. 704; [1949] 1 All E.R. 743, C.A.; *Braham* v. *J. Lyons & Co.* [1962] 1 W.L.R. 1048; [1962] 3 All E.R. 281, C.A.

Where the statutory duty is a duty of doing what is reasonably practicable, there may not be much difference between this standard and the common law standard of taking reasonable care. In one case the Court of Appeal have said that the standard in the two cases is much the same,[64] in another that the statutory obligation places a stricter obligation on the employer than the common law does.[65] Where an action is brought for breach of a statutory duty which is qualified by the words "reasonably practicable," in general, it is for the defendant to prove that he took action to meet the obligation imposed so far as was reasonably practicable. The plaintiff is under no obligation to prove a course of action that is reasonably practicable.[66]

(e) "Efficient"

Lighting can be efficient, that is, adequate to enable work to be done in safety, even though it may be poor.[67] If an object has to be maintained in an efficient state, that means efficient from a safety point of view.[68] "Efficiently lighted" means "reasonably well lighted for the purpose for which the light is there."[69] To be efficient the efficiency must be unqualified; thus it will not be sufficient to provide the most efficient guard practicable if the requirement is that the guard shall be efficient.[70]

(f) "Provided" and "available"

If a guard is provided, but the employees are forbidden to use it, there is no provision in law.[71] It has been said that while "providing" means "supplying" or "furnishing," a thing cannot be "provided" unless it is readily and obviously available.[72] Where an obligation was to have a piece of equipment "available," it was held that to have it at a place which was ten minutes' walk away was not a compliance.[73]

---

[64] *Jones* v. *N.C.B.* [1957] 2 Q.B. 55; [1957] 2 All E.R. 155, C.A.
[65] *Trott* v. *W. E. Smith* [1957] 1 W.L.R. 1154; [1957] 3 All E.R. 500, C.A.
[66] *Nimmo* v. *Alexander Cowan & Sons*, 1967 S.C., H.L. 79.
[67] *Kerridge* v. *P.L.A.* [1952] 2 Lloyd's Rep. 142.
[68] *Payne* v. *Weldless Steel* [1956] 1 Q.B. 196; [1955] 3 All E.R. 612, C.A.
[69] *Cowhig* v. *P.L.A.* [1956] 2 Lloyd's Rep. 306.
[70] *Vickers* v. *Gomme* [1957] 1 W.L.R. 656; [1957] 2 All E.R. 60, C.A.
[71] *Murray* v. *Schwachman* [1938] 1 K.B. 130; [1937] 2 All E.R. 68, C.A.
[72] *Norris* v. *Syndic* [1952] 2 Q.B. 135; [1952] 1 All E.R. 935, C.A. See also *Bux* v. *Slough Metals Ltd.* [1973] 1 W.L.R. 1358; [1974] 1 All E.R. 262, C.A.
[73] *Roberts* v. *Dorman Long & Co.* [1953] 1 W.L.R. 942; [1953] 2 All E.R. 428, C.A.

(g) Maintenance and repair

Where there is an obligation that something shall be "properly maintained," the duty is an absolute one.[74] But, although the duty is absolute, the word "maintain" in itself only refers to the general condition and soundness of construction of the object in question as, *e.g.* a floor),and does not include a duty in respect of some transient and exceptional condition, such as oil and water coming on to a floor due to a sudden storm.[75] There is, however, a duty to keep factory floors free from obstructions or substances likely to cause persons to slip.[76]

(3) PROOF

(a) Onus of proof

It has been decided by the House of Lords that the onus of proof is on the plaintiff to prove not only the breach of duty complained of, but also that such breach caused, or materially contributed to, his injury.[77] In this sense, a contribution is a material one unless so minimal as to be insignificant. But the rule that the plaintiff must prove both breach and causation is not one that is to be pressed too far; if, for instance, there is an obligation to provide a proper system of ventilation, and if it is proved that there was no system, or only an inadequate system, of ventilation, it requires little further to establish a causal link between the default and the illness due to noxious dust of a person employed in the shop.[78]

But where an obligation is contained in a proviso (*e.g.* "so far as is reasonably practicable"), the onus of proving compliance with the proviso is on the defendants.[79]

Proof of causation may be by direct evidence or by inference

---

[74] *Galashiels Gas Co.* v. *O'Donnell* [1949] A.C. 275; [1949] 1 All E.R. 319, H.L.; 1949 S.C., H.L. 31; *Payne* v. *Weldless Steel* [1956] 1 Q.B. 196; [1955] 3 All E.R. 612, C.A.; *Hamilton* v. *N.C.B.* [1960] A.C. 633; [1960] 1 All E.R. 76, H.L.; 1960 S.C., H.L. 1.
[75] *Latimer* v. *A.E.C.* [1953] A.C. 648; [1953] 2 All E.R. 449, H.L.
[76] Factories Act 1961, s.28.
[77] *Bonnington Castings* v. *Wardlaw* [1956] A.C. 613; [1956] 1 All E.R. 615, H.L.; 1956 S.C., H.L. 26.
[78] *Bonnington Castings* v. *Wardlaw, supra, per* Lord Reid; *Nicholson* v. *Atlas Steel Foundry and Engineering Co.* [1957] 1 W.L.R. 613; [1957] 1 All E.R. 776, H.L.; 1957 S.C., H.L. 44.
[79] *Marshall* v. *Gotham* [1954] A.C. 360; [1954] 1 All E.R. 937, H.L.

from proved facts. In either case the plaintiff must establish a balance of probability in his favour.[80]

Difficult questions sometimes arise where an employee contracts a disease which could have arisen from independent causes or which alternatively could have been caused by exposure to conditions of work for which the employer would be liable if there were proof that damage resulted therefrom. Lord Reid in the House of Lords put the test to be applied in such cases as follows:

> "When a man who has not previously suffered from a disease contracts that disease after being subjected to conditions likely to cause it, and when he shows that it starts in a way typical of disease caused by such conditions, he establishes a prima facie presumption that his disease was caused by those conditions."[81]

But, as Lord Reid went on to point out, the presumption can be displaced in many ways, such as by the employers showing that it was negatived by the subsequent course of the disease, or by proving some other cause as an equally probable cause of its origin.

(b) Whether plaintiff would have used safeguard if provided

Formerly there were a number of decisions which laid down the rule that, where there was a duty on a defendant to provide a safeguard, and he had failed to provide it, it did not lie in his mouth to say that the safeguard would have been of no use if he had provided it.[82] However, since the decision of the House of Lords in *Bonnington Castings* v. *Wardlaw*,[83] this rule certainly requires some modification, and indeed in one subsequent case[84] it was held that the claim of the plaintiff failed because he had failed to prove to the satisfaction of the court that he would have worn goggles if the employer in pursuance of his duty had in fact provided them. So also where there was failure to prove that if the employer had provided safety hooks, a window cleaning employee would have used them.[85] The principle has been further applied by the

---

[80] *Nicholson* v. *Atlas Steel Foundry and Engineering Co.*, note 78 *supra*.
[81] *Gardiner* v. *Motherwell Machinery & Scrap Co.* [1961] 1 W.L.R. 1424.
[82] *Bonham-Carter* v. *Hyde Park Hotel* [1948] W.N. 89; 64 T.L.R. 177; *Roberts* v. *Dorman Long & Co.* [1953] 1 W.L.R. 942; [1953] 2 All E.R. 428, C.A.; *Drummond* v. *British Building Cleaners* [1954] 1 W.L.R. 1434; [1954] 3 All E.R. 507, C.A.
[83] [1956] A.C. 613, H.L.
[84] *Nolan* v. *Dental Manufacturing Co.* [1958] 1 W.L.R. 936; [1958] 2 All E.R. 449.
[85] *Wigley* v. *British Vinegars* [1962] 3 W.L.R. 731; [1962] 3 All E.R. 161, H.L.

House of Lords; the plaintiff failed to prove that an employee steel erector would have used a safety belt if one had been provided.[86]

(4) DEFENCES

(a) Contracting out

A person under a statutory duty cannot by agreement with the plaintiff contract out of that duty.[87]

(b) Delegation to plaintiff

There is some authority for saying that, if the defendants can establish that they validly delegated the performance of their statutory duty to the plaintiff himself, they will not be liable if the plaintiff, in attempted performance of that duty, causes a breach of statutory duty to arise.[88] But it may be doubtful whether such a defence really exists, except in cases where the statutory provisions expressly provide for the delegation of the duty in whole or in part to the plaintiff.[89] If delegation is a separate defence on its own, it can only apply to the delegation of a positive, rather than a negative, duty,[90] and in addition, to constitute delegation, there must be clear evidence of express delegation; merely pointing out to an employee that he must comply with the regulation does not constitute delegation.[91]

It has, however, been said that the true test in such cases is not to ask whether there has been delegation, but "to ask the usual question: whose fault was it?"[92]

---

[86] *McWilliams* v. *Sir William Arrol & Co. Ltd.* [1962] 1 W.L.R. 295; overruling *Roberts* v. *Dorman Long & Co.* [1953] 1 W.L.R. 942. This case was distinguished by Lord Guest in *Ross* v. *Associated Portland Cement Manufacturers, Ltd.* [1964] 1 W.L.R. 768. *Cf. Money* v. *Thorn Electrical Industries* [1977] 11 C.L. (unreported), C.A.
[87] *Lavender* v. *Diamints* [1949] 1 K.B. 585; [1949] 1 All E.R. 532, C.A.
[88] *Smith* v. *Baveystock* [1945] 1 All E.R. 531, C.A.; *Barcock* v. *Brighton Corporation* [1949] 1 K.B. 339; [1949] 1 All E.R. 251; *Johnson* v. *Croggan & Co.* [1954] 1 W.L.R. 195; [1954] 1 All E.R. 121.
[89] As in *Smith* v. *Baveystock, supra.*
[90] *Gallagher* v. *Dorman Long* [1947] 2 All E.R. 38; 177 L.T. 143, C.A.
[91] *Manwaring* v. *Billington* [1952] W.N. 467; [1952] 2 All E.R. 747, C.A.
[92] *Ginty* v. *Belmont Building Supplies* [1959] 1 All E.R. 414.

(c) Delegation to third party

If a defendant undertakes work which imposes on him a statutory duty, he cannot divest himself of that duty by sub-contracting,[93] nor by relying on the fact that the object in question is under the control of a third party.[94] Even if a third party gives assurances to a person under a statutory duty that they have performed the duty of the latter, the latter will still be liable if the assurances are incorrect.[95]

(d) Act of independent contractor

In general, act of independent contractor is no defence to a breach of statutory duty.[96]

(e) Latent defect

Latent defect is no defence to an absolute statutory obligation.[97] Whether it constitutes a defence in cases where the statutory duty is qualified depends on the words of qualification used.

(f) Breach of employer caused solely by act of plaintiff

A question arises when a breach of duty by an employer results solely from something done or omitted by the plaintiff himself, such as where the plaintiff deliberately removes a guard from a properly guarded machine.[98]

In this connection a test of general application has been laid down in the following words:

> "The deceased and the defendants are both in breach of statutory duty . . . and the defendants' breach was brought about

---

[93] *Mulready* v. *Bell* [1953] 2 Q.B. 117; [1953] 2 All E.R. 215, C.A.
[94] *Heard* v. *Brymbo Steel Co.* [1947] K.B. 692; 177 L.T. 251, C.A.
[95] *Jerred* v. *Dent* [1948] 2 All E.R. 104; 81 Ll.L.R. 412.
[96] *Hosking* v. *De Havilland Aircraft Co.* [1949] 1 All E.R. 540; 83 Ll.L.Rep. 11; *Mulready* v. *Bell* [1953] 2 Q.B. 117; [1953] 2 All E.R. 215, C.A.; *Dooley* v. *Cammell Laird* [1951] 1 Lloyd's Rep. 271.
[97] *Edwards* v. *N.C.B.* [1949] 1 K.B. 704; [1949] 1 All E.R. 743, C.A.; *Whitehead* v. *James Stott* [1949] 1 K.B. 358; [1949] 1 All E.R. 245, C.A.; *Galashiels Gas Co.* v. *O'Donnell* [1949] A.C. 275; [1949] 1 All E.R. 319, H.L.; 1949 S.C.(H.L.) 31; *Hamilton* v. *N.C.B.* [1960] A.C. 633; [1960] 1 All E.R. 76, H.L.; 1960 S.C., H.L. 1.
[98] *Norris* v. *Syndic* [1952] 2 Q.B. 135; [1952] 1 All E.R. 935, C.A., in which case the defendants were held 80 per cent. to blame, owing to their own bad example in tolerating a practice of removing guards.

by the deceased's breach; that is to say, the deceased . . . at once committed a breach of his own statutory duty and put the defendants in breach of theirs. In these circumstances I cannot for my part see how the plaintiff can succeed merely on proof of the defendants' breach of statutory duty. . . . In my view it must be necessary for the plaintiff . . . to go on and prove that the breach of statutory duty . . . was in some degree due to the defendants' negligence over and above their statutory responsibility for a mere innocent breach of the regulations. The inquiry on which one must embark thus does not differ greatly from the inquiry necessary in a case where the issue is one of negligence or no negligence at common law."[99]

In *Ross* v. *Associated Portland Cement Manufacturers, Ltd.*[1] a workman was killed when he fell from the top of a long ladder resting against some wire netting which he and others were mending. The employers were guilty of a breach of statutory duty in not providing a movable platform. It was held that the employers could not entirely escape liability, despite the workman's own negligence, since they had failed to provide proper equipment, had not kept the place of work safe, and by their failure to provide suitable equipment had brought the workman to decide to use an unsatisfactory ladder. But different considerations arise where the act or omission of the plaintiff which gives rise to the breach is outside his competence. In such a case liability for the breach may well fall wholly on the employer.[2] Once the plaintiff has established that there was a breach of an enactment which made the employer absolutely liable, and that that breach caused the accident, he need do no more. But it is then open to the employer to set up a defence that in fact he was not in any way in fault, but that the plaintiff employee was alone to blame. To do this, it is not sufficient that the employer has been exonerated from any common law negligence since statutory obligation may exceed common law obligation.[3] Further, whilst there may be no need to instruct a skilled employee about obvious dangers, there is a need to instruct him about the application of regulations, where no danger is apparent.[4]

---

[99] *Davison* v. *Apex Scaffolds* [1956] 1 Q.B. 551; [1956] 1 All E.R. 473, C.A., *per* Jenkins L.J.
[1] [1964] 2 All E.R. 452, H.L. See also *Donaghey* v. *P. O'Brien & Co.* [1966] 1 W.L.R. 1170.
[2] *Byers* v. *Head Wrightson* [1961] 1 W.L.R. 961; [1961] 2 All E.R. 538.
[3] *Boyle* v. *Kodak Ltd.* [1969] 1 W.L.R. 661.
[4] *Boden* v. *Moore* (1961) 105 S.J. 510, C.A.

(g) Contributory negligence

The principles applying to contributory negligence as a defence have already been dealt with elsewhere.[5] It has been said that the actions of workmen must not be judged too severely in determining whether carelessness amounts to contributory negligence in an action for breach of statutory duty, for too strict a standard would defeat the object of regulations.[6] It is, however, possible in an appropriate case for a finding to be made of 100 per cent. contributory negligence on the plaintiff's part, thus negating any award of damages, even if the intention of the statutory provision is to provide against folly on the part of the employee.[7]

(h) Consent to the risk ("volenti non fit injuria")

It would now seem, as a result of the decision of the House of Lords in *Imperial Chemical Industries, Ltd.* v. *Shatwell*[8] that in cases of injury arising from a breach of statutory duty, it may be possible for an employer to raise the defence of consent, either express or implied, to run the risks involved in carrying on an operation even though the necessary statutory safeguards have not been taken. This will be possible where, as in the case in question, the duty is imposed by statute directly on the injured workman himself, and not on the employer, so that the employer is only vicariously responsible for a breach of the duty by reason of the workman's failure to observe it. In that case the plaintiff and his brother were guilty of a breach of the Quarries (Explosives) Regulations, 1959, in testing a circuit connected with explosives without withdrawing to a place of safety and taking shelter. Both men knew the dangers involved in this conduct. It was held by the House of Lords that the conduct of the plaintiff's brother, in collaborating in the forbidden and unlawful method of conducting the test, was causally connected with the plaintiff's injuries, and that *volenti non fit injuria* was a complete defence where, as here, two

---

[5] pp. 34–35 above.
[6] *Per* Sachs L.J. in *Mullard* v. *Ben Line Steamers Ltd.* [1970] 1 W.L.R. 1414; [1971] 2 All E.R. 424; [1970] 2 Lloyd's Rep. 121, C.A. *Cf. Thornton* v. *Swan Hunter (Shipbuilders) Ltd.* [1972] 2 Lloyd's Rep. 112, C.A., where it was said that an employer was not stopped from raising the defence of contributory negligence to an action for breach of a strict statutory duty, s.14(1) of the Factories Act 1961, where the plaintiff employee's action created the breach of duty.
[7] *Jayes* v. *IMI (Kynoch) Ltd.* [1985] I.C.R. 155, C.A. (breach of reg. 5 of the Operations at Unfenced Machinery Regulations 1938).
[8] [1965] A.C. 656, H.L., followed in *Bolt* v. *Moss and Sons Ltd.* (1966) 110 S.J. 385.

fellow-servants combined to disobey an order deliberately though they knew the risk involved.

However, the general rules of law about the defence of *volenti non fit injuria*, including the Unfair Contract Terms Act 1977, will still be applicable.[8a]

(5) DAMAGES

The principles applied in awarding damages for personal injuries caused by breach of statutory duty are the same as those where the employer is in breach of his common law duty to take reasonable care for the safety of his workmen.[9]

The requirement of compulsory insurance against liability for personal injury applies in respect of personal injuries caused by breach of statutory duty as it applies in respect of personal injuries caused by breach of duty owed at common law.[10]

[8a] See pp. 36–37.
[9] See Chap. 2.
[10] See pp. 55–56.

4. Liability of Employer as Occupier

(a) Introductory

The Occupiers' Liability Act, 1957, affects the occupiers of all premises, not merely those occupying factories, shops and offices. Since there may be circumstances in which an employer who is an occupier of premises may be involved in liability to "visitors" thereon, arising out of the state of, or the way operations are conducted in the factory or other premises concerned, some account of the provisions of this Act and the way they have been interpreted, may be considered useful. So far as concerns liability to those who are not "visitors," *e.g.* trespassers, the law is to be found in the Occupiers' Liability Act 1984.[1]

This Act, which replaced the previous common law on the subject of the liability of an occupier to persons entering the premises in his occupation,[2] lays down a general duty of care, called the "common duty of care," which is owed by an occupier to all lawful visitors. It in no way affects the employer's liability to his employees and others under the Health and Safety at Work Act 1974, the Factories Act, 1961, or the Offices, Shops and Railway Premises Act, 1963.

(b) Scope of the 1957 Act

The Act regulates the duty which an occupier of premises owes to his visitors in respect of dangers due to the state of the premises or to things done or omitted to be done on them. This duty is imposed by the law in consequence of a person's occupation or control of premises.[3] Under the Act, the same obligations are owed where a person is in occupation or control of any fixed or

---

[1] 1984, c. 3. See pp. 84–85.
[2] Occupiers' Liability Act 1957 (5 & 6 Eliz. 2, c. 31), s.1(1): except as regards who are occupiers and who are visitors: see below.
[3] *Ibid.*, s.1(2).

movable structure, including any vessel, vehicle or aircraft.[4] Moreover, the same obligations are owed in respect of damage to property, as well as to the person, including the property of persons who are not themselves visitors on the premises.[5]

(c) Who is an occupier

There is no definition in the Act of what is meant by occupation of premises or structures for the purposes of the Act. Indeed it is specifically stated that reference must be made to the common law to establish the meaning of these expressions.[6] It is to be noted, however, that, under the Act,[7] by a provision which is probably intended to apply in cases of occupation by a tenant of residential premises (though it would seem also to be applicable where the premises are a factory, etc.), a landlord out of occupation may be under the obligations imposed by the Act, where the landlord, by virtue of the tenancy, is obliged to maintain and repair the premises. In such instances, the landlord will be treated as an occupier of the premises, for the purposes of the obligations under the Act, so far as concerns dangers arising out of any default by the landlord in carrying out his duty as regards maintenance and repair.

So far as the meaning of occupation is concerned, it has been held that this includes not only actual control but also control exercised through another. In *Bunker* v. *Charles Brand Ltd.*[8] O'Connor J. held that the defendants, main contractors for the construction of the Victoria Underground line, remained occupiers of a tunnel and the digger in it, even though they had engaged the plaintiff's employers, specialist contractors, to perform modifications to it. Two persons may be in control of premises for the purposes of the Occupiers' Liability Act, 1957.[9]

(d) Who are visitors

The Act does not alter the rules of the common law as to the persons to whom the duty imposed by the Act is owed.[10] Accordingly the persons who are to be treated as visitors of an occupier are the

---

[4] *Ibid.*, s.1(3)(*a*).
[5] *Ibid.*, s.1(3)(*b*).
[6] *Ibid.*, s.1(2).
[7] *Ibid.*, s.4.
[8] [1969] 2 W.L.R. 1392; [1965] 2 All E.R. 59.
[9] *Fisher* v. *C.H.T. Ltd.* [1966] 1 All E.R. 88.
[10] Occupiers' Liability Act, 1957, s.1(2).

same as the persons who would at common law be treated as an occupier's invitees or licensees.[11] This means that the Act extends its protection to all lawful visitors of an occupier, whether they come on the premises for their own or the occupier's benefit, as long as they do so with his permission, or in the exercise of a right conferred by law, whether with the occupier's permission or not, *e.g.* factory inspectors.[12] There is no need now, as there was before the Act, to differentiate invitees from licensees. By virtue of another provision of the Act,[13] where a person enters or uses premises under a contract which gives him the right to do so, or brings or sends goods to such premises, he will be in the same position as any other lawful visitor, except to the extent to which the contract imposes a stricter duty on the occupier than would be imposed by the Act.

The duty of an occupier to a person who is not a "visitor," *e.g.* a trespasser is governed by the Occupiers' Liability Act 1984. (see below).

(e) The duty owed

The occupier owes a common duty of care to all his visitors.[14] This is defined as a duty to take such care as in all the circumstances of the case is reasonable to see that the visitor will be reasonably safe in using the premises for the purposes for which he is invited or permitted by the occupier to be there.[15] Hence, no duty of care is owed under the Act wherever and whenever a visitor is not acting

---

[11] *Ibid.*
[12] *Ibid.*, s.2(6). But this does not extend to persons exercising a public or private right of way over premises: *Greenhalgh* v. *British Railways Board* [1969] 2 W.L.R. 892; [1969] 2 All E.R. 114, C.A.
[13] *Ibid.*, s.5(1). This applies to fixed and movable structures: *ibid.*, s.5(2). But it does not apply to contracts for the hire of, or for the carriage for reward of persons or goods in, any vehicle, vessel, aircraft, or other means of transport, or by virtue of any contract of bailment: *ibid.*, s.5(3). S.5(1) does not preclude a person who enters premises under a contract with the occupier from making a claim in tort as a visitor under s.2(1): *Sole* v. *W. J. Hallt Ltd.* [1973] 1 Q.B. 574. This is material where, as in that case, the plaintiff's contributory negligence was such as to amount to a *novus actus interveniens* and a break in the chain of causation which defeated a claim in contract, but only reduced the damages awarded on the claim in tort.
[14] *Ibid.*, s.2(1).
[15] *Ibid.*, s.2(2). See *Cook* v. *Broderip* (1968) 112 S.J. 193, following *Green* v. *Fibreglass Ltd.* [1958] 2 Q.B. 245: employer not liable as occupier when electrician engaged to do work on circuit which later resulted in injury to employee. Thus permission may be given to enter only if the premises are lit, which may affect the occupier's duty: see *Hogan* v. *P. & O. Steam Navigation Co.* [1959] 2 Lloyd's Rep. 305.

in pursuance of his invitation or permission. The duty is only to take reasonable care (which will be a question of fact in all cases). It is not to guarantee the safety of the visitor.

In deciding what would be reasonable care in all the circumstances of the case, it is relevant to consider the degree of care and of want of care which would ordinarily be looked for in the kind of visitor involved.[16] For example, in proper cases an occupier must be prepared for children to be less careful than adults.[17]

The Act also gives as an example of what may be relevant that an occupier may expect that a person, in the exercise of his calling, will appreciate and guard against any special risks ordinarily incidental to it, so far as the occupier leaves him free to do so.[18]

(f) Exclusion of duty

Under the Act, the occupier may be free to, and may extend, restrict, modify or exclude his duty to any visitor or visitors by agreement or otherwise.[19] However, by the Unfair Contract Terms Act 1977,[20] it is not possible for a person carrying on a business to exclude or restrict liability by contract made on or after February 1, 1978 (or by any notice) for negligently caused:

(i) death or personal injury *or*
(ii) any other kind of damage or loss, unless in this latter case, the exclusion or restriction is proved to be reasonable.[21]

These provisions are expressly made applicable to the common duty of care under the Occupiers' Liability Act 1957[22] but the Occupiers' Liability Act 1984 provides that "liability of an occu-

---

[16] *Ibid.*, s.2(3). This is distinct from the possibility that, although an occupier may owe the common duty of care, the visitor may be guilty of contributory negligence. See *Sandford* v. *Eugene Ltd.* (1971) 115 S.J. 33.
[17] Occupiers' Liability Act, 1957, s.2(3)(*a*).
[18] Occupiers' Liability Act 1957, s.2(3)(*b*). This did not apply to a place or means of access to a place: *per* Lord Denning M.R., *Woollins* v. *British Celanese* (1966) 110 S.J. 686. A shipowner is not required to notify stevedores unloading his ship, or their employer, of the method used to load the ship, where it is a method that is commonly used: *Crawley* v. *Gracechurch Line Shipping Ltd.* [1971] 2 Lloyd's Rep. 179.
[19] Occupiers' Liability Act, 1957, s.2(1). But the occupier is not free to restrict or exclude the common duty of care as regards visitors entering in pursuance of a contract between his employer and the occupier: *ibid.*, s.3(1). The occupier must perform his obligations under such contract in so far as those obligations go beyond the common duty of care: *ibid.*
[20] 1977, c. 50, ss.1–14.
[21] Defined in s.11.
[22] s.1(1)(*c*).

pier of premises for breach of an obligation or duty towards a person obtaining access to the premises for recreational or educational purposes, being liability for loss or damage suffered by reason of the dangerous state of the premises, is not a business liability of the occupier unless granting that person such access for the purposes concerned falls within the business purposes of the occupier."[23] A notice purporting to exclude or restrict the liability of an occupier under the 1957 Act will have effect only in so far as it is not nullified by the above provisions of the Unfair Contract Terms Act 1977 and, even then, the 1977 Act provides that so far as a contract term or notice is concerned, "a person's agreement to or awareness of it is not of itself to be taken as indicating his voluntary acceptance of any risk."[24]

Moreover, the common duty of care does not impose on the occupier any obligation to a visitor in respect of risks willingly accepted as his by the visitor.[25] The question whether a risk was so accepted is to be decided on the same principles as in any other case in which one person owes a duty of care to another.[26]

(g) Performance of duty

In determining whether an occupier has properly performed his common duty of care, regard is to be had to all the circumstances.[27] In this respect a warning to the visitor may be enough to amount to a proper performance of his duty by the occupier. This will only result, however, where, in all the circumstances, the warning was enough to enable the visitor to be reasonably safe.[28] Thus in *Roles* v. *Nathan*[29] the majority of the Court of Appeal held that the warnings given to chimney sweeps of the danger of gas which in the event caused their death was enough to enable them to be reasonably safe, if they had heeded the warnings.

Delegation to an independent contractor of the task of making premises or structures reasonably safe may also be an adequate

---

[23] 1984, c. 2, s.2, adding the cited words to section 1(3) of the Unfair Contract Terms Act 1977.
[24] s.2(3). *Cf. Ashdown* v. *Williams and Son Ltd.* [1957] 1 Q.B. 409, C.A.
[25] Occupiers' Liability Act, 1957, s.2(5).
[26] Occupiers' Liability Act, 1957, s.2(5). See *Bunker* v. *Charles Brand Ltd.* [1969] 2 W.L.R. 1392; [1969] 2 All E.R. 59. See now, however, the provision of s.2(3) of the Unfair Contract Terms Act 1977 referred to above.
[27] Occupiers' Liability Act, 1957, s.2(4).
[28] *Ibid.*, s.2(4)(*a*).
[29] [1963] 2 All E.R. 908, C.A. *Cf. Bishop* v. *J. S. Starnes & Sons Ltd. and MacAndrews & Co. Ltd.* [1971] 1 Lloyd's Rep. 162, where it was recognised that there may be a difference between the degree of warning required from an occupier who is also the employer and that required from an occupier who is not.

discharge by the occupier of his duty under the Act. Where damage is caused to a visitor by a danger due to the faulty execution of any work of construction, maintenance or repair by an independent contractor employed by the occupier, the occupier will not be treated as answerable for the danger if in all the circumstances he acted reasonably in entrusting the work to an independent contractor and took such steps as he reasonably ought to satisfy himself that the contractor was competent and the work had been properly done.[30] In other words, in the absence of personal negligence on the part of the occupier, he will not be liable where damage results from negligent work by his independent contractor which makes premises unsafe.

(h) Application to Crown

The 1957 Act binds the Crown, but only in so far as the Crown may be made liable in tort by the Crown Proceedings Act, 1947.[31]

(i) Liability to non-visitors, e.g. trespassers

The liability of the occupier to a person not a visitor, *i.e.* normally a trespasser, is governed by the Occupiers' Liability Act 1984,[32] which does not, however, apply to the highway[33] or to any loss of or damage to property, as distinct from personal injuries.[34] The 1984 Act imposes a duty on the occupier of premises[35] towards non-visitors to take such case as is reasonable in all the circumstances of the case to see that they do not suffer injury on the premises[36] from dangers due to the state of the premises[37] or to things done or omitted to be done on them.[38] However, the occupier's liability extends only if:

---

[30] Occupiers' Liability Act, 1957, s.2(4)(*b*).
[31] *Ibid.*, s.6. See pp. 7–8.
[32] 1984, c. 3. As to the law prior to the coming into force of the 1984 Act, see *Cooke* v. *Midland G.W.Ry. of Ireland* [1909] A.C. 229; *Addie & Sons Ltd.* v. *Dumbreck* [1929] A.C. 358; *Excelsior Wire Rope Co.* v. *Callan* [1930] A.C. 404; and *British Railways Board* v. *Herrington* [1972] A.C. 877.
[33] 1984 Act, s.1(7)—"highway" does not include a ferry or waterway—s.1(9).
[34] "Injury" means "anything resulting in death or personal injury, including any disease and any impairment of physical or mental condition"—s.1(9).
[35] "Premises" include any fixed or movable structure (including any vessel, vehicle or aircraft)—s.1(2) and (9).
[36] *Ibid.*
[37] *Ibid.*
[38] 1984 Act, s.1(1)(*a*).

(a) he is aware of the danger or has reasonable grounds to believe that it exists *and*
(b) he knows or has reasonable grounds to believe that the non-visitor is in the vicinity of the danger concerned or that he may come into the vicinity of the danger (in either case, whether or not the non-visitor has lawful authority for being in that vicinity) *and*
(c) the risk is one against which, in all the circumstances of the case, the occupier may reasonably be expected to offer the non-visitor some protection.[39]

The duty may in an appropriate case be discharged by taking such steps as are reasonable in all the circumstances of the case to give warning of the danger concerned or to discourage persons from incurring the risk.[40] Moreover, no duty is owed in respect of risks willingly accepted as his by the non-visitor.[41] The 1984 Act binds the Crown.[42]

(j) Disabled persons

Under section 8A of the Chronically Sick and Disabled Persons Act 1970 (c. 44)[43] any person undertaking the provision of factory, office, shop or railway premises shall in the means of access both to and within the premises, and in the parking facilities and sanitary conveniences to be available (if any), make provision, in so far as it is in the circumstances both practicable and reasonable, for the needs of persons using the premises who are disabled.

No sanction is imposed by the 1970 Act for non-compliance with section 8A, but it might be that breach of the section could ground a civil action for damages for breach of statutory duty.

---

[39] 1984 Act, s.1(3).
[40] 1984 Act, s.1(5).
[41] 1984 Act, s.1(6)—the question whether a risk was so accepted is to be decided on common-law principles (*ibid.*).
[42] 1984 Act, s.3, "but as regards the Crown's liability in tort shall not bind the Crown further than the Crown is made liable in tort by the Crown Proceedings Act 1947"—(*ibid.*).
[43] Inserted by the Chronically Sick and Disabled Persons (Amendment) Act 1976 (c. 49) and effective from October 29, 1976.

5. Liability for Fatal Accidents

(1) INTRODUCTORY

(a) Common law

Originally, the common law permitted recovery of damages on the ground of negligence or breach of statutory duty, only where such wrongs caused injury. If the injured employee died as a consequence of his injuries—or from any other cause died before he could obtain judgment against the employer—his cause of action died with him. To this there were exceptions, none of which were capable of providing the deceased employee's estate with a remedy. Nor did the killing of an employee, as a result of negligence or breach of statutory duty, give rise to any liability to a third person, *e.g.* the dependants of the employee.

(b) Statutory change

Both these bars to recovery where an employee died from injuries received at work as a result of his employer's negligence or breach of statutory duty, have been removed by statute, curiously enough, the first to be so removed being the prohibition of an action by the employee's dependants. The law governing the situation where a fatal accident has occurred at work is now contained in the Law Reform (Miscellaneous Provisions) Act 1934, and the Fatal Accidents Act 1976, which consolidated the Fatal Accidents Acts 1846–1959 in respect of deaths occurring on or after September 1, 1976.

(2) SURVIVAL OF CAUSES OF ACTION

(a) The provisions of the Act

The Law Reform (Miscellaneous Provisions) Act 1934 (s.1) provides that on the death of any person all causes of action vested in him survive for the benefit of his estate. Thus there must be a right to sue vested in the deceased employee at the time of his death if his estate is to be able to pursue such right after his death. The Act does not create any new cause of action. It simply retains in existence, so that the estate of the deceased employee may benefit, any cause of action which arose prior to his death. Where the accident on which liability is to be founded results in the immediate death of the employee, his cause of action is treated as having vested the instant before his death, so as to permit its being regarded as vested and subsisting at the time of death.

(b) Cause of action

It follows from the provisions referred to above that, if there would have been a valid defence capable of being raised against the employee himself, had he not died but had brought an action on his own behalf, any such defence may be raised in an action brought by the estate of the deceased employee under the Act of 1934. Thus, the plea that the employee consented to the risk[1-2] and the defence of contributory negligence,[3-4] if applicable, may be raised by the employer in an action brought under this statute in the same way as they could be raised in an action brought by the injured employee.

(c) Damages recoverable

Where the death has been caused by the act or omission which gives rise to the cause of action, the damages must be calculated without reference to any loss or gain to his estate consequent on the death.[5] Thus if, *e.g.* an employee would have inherited a large sum of money had he survived, the loss of this sum because of his death is irrecoverable in an action under this Act. On the other hand, if by his death his estate succeeds to a sum of money, *e.g.* under a policy of insurance, that gain must also be disregarded.[6]

[1-2] pp. 34–35.
[3-4] pp. 36–37.
[5] 1934 Act, s.1(2)(*c*).
[6] But rights under the Fatal Accidents Act 1976, are not affected by this provision; they are conferred upon the dependants of a deceased person, not on his estate: Law Reform (Miscellaneous Provisions) Act 1934, s.1(5). Neither is loss of future earnings to be regarded as a gain to the estate and an award can properly be made under this head—*Kandalla* v. *British Airways Corp.* (1979) 123 S.J. 769.

However, funeral expenses, incurred by the estate because of the death may be recovered from the employer.[7]

Damages may also be recovered in respect of the following kinds of loss:

(a) medical and hospital expenses, if incurred;
(b) pain and suffering;
(c) loss of earnings and profits up to the date of death.[8]

Before the Administration of Justice Act 1982[9] damages could include:

(a) a conventional sum of approximately £1,250 for loss of expectation of life, even though the deceased was killed instantaneously[10] and
(b) damages for loss by the deceased of income he would have earned had his expectation of life not been diminished.[11]

But the 1982 Act has abolished[12] head of damage (a) and has provided[13] that head of damage, (b) though recoverable by a living plaintiff, cannot be recovered in a claim under the 1934 Act.

Moreover, under the Act of 1934 (as amended by the 1982 Act)[14] damages under the 1934 Act cannot include (a) any exemplary damages; (b) a claim for bereavement damages that the deceased himself had in respect of the death of a child or spouse, under the Fatal Accidents Act 1976 (as amended by the 1982 Act).[15]

(3) LIABILITY TO DEPENDANTS

(a) Alteration of the common law

Originally the infliction of death by some wrongful act, *e.g.*, negligence or breach of statutory duty, not only extinguished any cause of action which the deceased might have pursued in the courts, it also gave rise to no liability, either to the estate of the deceased or

---

[7] *Ibid.* s.1(2)(*c*).
[8] Subject to deductions for income tax payable in respect of such earnings: *British Transport Commission* v. *Gourlay* [1956] A.C. 185.
[9] 1982, c. 53.
[10] *Rose* v. *Ford* [1937] A.C. 826, H.L.; *Benham* v. *Gambling* [1941] A.C. 157, H.L.
[11] *Gammell* v. *Wilson* [1981] 2 W.L.R. 248, H.L.
[12] 1982 Act, s.1(1)(*a*).
[13] 1982 Act, s.4(2).
[14] 1982 Act, s.4.
[15] 1982 Act, s.3(1).

to any other person or persons injuriously affected by his death. This was changed by the Fatal Accidents Act, 1846, section 1 (now replaced by s.1 of the Fatal Accidents Act 1976 (as substituted by s.3 of the Administration of Justice Act 1982) providing that if the death of a person is caused by wrongful act, neglect or default, and the act, neglect or default is such as would (if death had not ensued) have entitled the party injured to maintain an action and recover damages in respect thereof, then the person who would have been liable if death had not ensued will be liable to an action for damages, notwithstanding the death of the person injured. But the statute did not give any remedy to the estate of the deceased person. It created an entirely new cause of action in favour of certain dependants of the deceased.[16] Consequently, an action under the Fatal Accidents Act 1976, is a separate action from any action which may be brought by the estate of a deceased employee under the Law Reform (Miscellaneous Provisions) Act 1934. Such actions may be brought distinctly from each other, though the beneficiaries may be the same under both actions, in which event certain consequences flow, as regards damages—see below—"Damages recoverable."

(b) By and for whom the action may be brought

The action under the Fatal Accidents Act 1976 must be brought by and in the name of the executor or administrator of the deceased employee.[17] If there is no executor or administrator, or the executor or administrator of the deceased employee fails to bring an action under the Act within six months of the death of the deceased, the action may be brought by and in the name of any person or persons for whose benefit the executor or administrator could have brought the action.[18] In such an action, as in an action brought by the executor or administrator, the benefits obtained are obtained for all those entitled under the Act,[19] but the fact that one dependant by his tort caused the death of the deceased (and therefore cannot claim under the Act) does not prevent another

---

[16] *Seward* v. *The Vera Cruz* (1884) 10 App.Cas. 59.
[17] 1976 Act, s.2, as substituted by s.3 of the Administration of Justice Act 1982 (c. 53). An action commenced by dependant(s) in their own right before the expiry of six months from the death is premature but the irregularity is cured by the expiry of the six months without the executor or administrator having brought an action and no fresh action by the dependant(s) is necessary—*Austin* v. *Hart* [1983] 2 All E.R. 341, P.C.
[18] *Ibid.*
[19] 1976 Act, s.3(1), as substituted by s.3 of the Administration of Justice Act 1982.

dependant from claiming under the Act against the tortious dependant.[20]

The following relatives[21] of the deceased are the beneficiaries on whose behalf an action may be brought under the Act[22]: wife, husband, or former wife or husband[23]; "common-law" husband or wife[24]; parent or any descendant[25]; child of any descendant[26]; brother or sister, uncle or aunt, or their issue.

(c) When the action is to be brought

The wrong committed by the defendant must amount to a tort in respect of which the employee could have sued had he lived. Thus if for some reason the employee could not have sued, *e.g.*, if he had settled the claim before his death,[27] or excluded the liability of

---

[20] *Dodds* v. *Dodds, The Times,* July 29, 1977.
[21] Including relationships by affinity and of the half-blood—Fatal Accidents Act 1976, (c. 30), s.1(5), as substituted by s.3 of the Administration of Justice Act 1982.
[22] Fatal Accidents Act 1976 (c. 30), s.1(3), as substituted by s.3 of the Administration of Justice Act 1982.
[23] *i.e.* where the marriage has been dissolved or annulled—Fatal Accidents Act 1976 (c. 30), s.1(3) and (4), as substituted by s.3 of the Administration of Justice Act 1982 (c. 53). *Cf. Davies* v. *Taylor* [1972] 3 W.L.R. 801; [1972] 3 All E.R. 836, H.L., as to separated spouses.
[24] *i.e.* not married but living with the deceased as husband or wife at the date of death and for two years previously—Fatal Accidents Act 1976 (c. 30), s.1(3), as substituted by s.3 of the Administration of Justice Act 1982.
[25] "Parent" includes a person treated by the deceased as his parent—Fatal Accidents Act 1976 (c. 30), s.1(3), as substituted by s.3 of the Administration of Justice Act 1982 (c. 53).
[26] "Child" includes:

(i) illegitimate, adopted, and step-children—Fatal Accidents Act 1976 (c. 30), s.1(3) and (5), as substituted by the Administration of Justice Act 1982;

(ii) posthumous children—*The George and Richard* (1871) L.R. 3 A. & E. 466;

(iii) any person treated by the deceased as a child of the family (of deceased's marriage)—Fatal Accidents Act 1976, s.1(3), as substituted by the Administration of Justice Act 1982.

[27] *Read* v. *Great Eastern Ry.* (1868) L.R. 3 Q.B. 555. In *McCann* v. *Sheppard* [1973] 1 W.L.R. 540, C.A., where the Court of Appeal reduced the damages awarded for loss of future earnings to a plaintiff who died after the trial, it was doubted whether his widow might then be able to bring an action under the Fatal Accidents Acts. *Per* James L.J., s.1 of the 1846 Act appears to put insuperable difficulties in the way of such proceedings.

the defendant,[28] no action under the Act will be available.[29] If the deceased employee was guilty of contributory negligence then the extent to which such conduct would have reduced the liability of the defendant to the deceased must be taken into account in determining what is recoverable in an action under the Fatal Accidents Act 1976.[30] But a claim under the Act is distinct from any other cause of action, *e.g.* one brought by the estate of the deceased under the Law Reform (Miscellaneous Provisions) Act, 1934. Hence periods of limitation are different.[31] Claims under the Fatal Accidents Act are subject to two limitation periods. First, no claim under the Act may be brought (subject to the court's overriding power—see below) if the deceased's own right of action was barred at death; that is if he died more than three years after the cause of action accrued to him or after his later "knowledge"[32] that he had a "significant injury."[33] Secondly, the action may not (subject to the court's overriding power—see below) be commenced more than three years after the deceased died or the later "knowledge" of the person for whose benefit the action is brought.[34] However, the court is given power to override both of these limitation periods, having regard to a number of matters specified in the Act, *e.g.* the reason for the delay, whether the defendant contributed to it, and the nature of any medical or legal advice received by the plaintiff.[35]

An action on behalf of the estate of the deceased under the Law Reform (Miscellaneous Provisions) Act 1934, is statute-barred three years after the date of death or the later "knowledge" of the deceased's personal representative, but this period also is subject to the overriding power of the court under the Limitation Act 1975.[36]

Even where the action is commenced in time, delay thereafter may cause the action to be dismissed for want of prosecution.[37]

---

[28] But an agreement limiting the liability of the defendant will not affect an action under the Fatal Accidents Acts: *Nunan* v. *Southern Ry.* [1924] 1 K.B. 223.
[29] The rule is now given statutory expression in Limitation Act 1980, s.12(1).
[30] 1976 Act, s.5.
[31] See *British Columbia Electric Ry. Co.* v. *Gentile* [1914] A.C. 1034.
[32] Limitation Act 1980, s.11.
[33] *Ibid.*
[34] Limitation Act 1980, s.11. As to the detailed definition of "knowledge," see pp. 52–53. Where some but not all the dependants have "knowledge" the action can be brought but the dependants who have knowledge will normally be debarred from sharing in its fruits (1980 Act, s.13).
[35] Limitation Act 1980, s.33.
[36] *Ibid.*
[37] See pp. 54–55.

(d) Damages recoverable

Under the Fatal Accidents Act 1976,[38] damages are awarded in proportion to the injury resulting from the death of the deceased to the parties respectively for whom and for whose benefit the action is brought. Such damages must be divided amongst the beneficiaries for whom the action is brought, in such shares as the court may decide.[39] In assessing damages under the 1976 Act, benefits which have accrued or will or may accrue to any person from the estate of the deceased or otherwise as a result of the death must be disregarded.[40] Funeral expenses may be recovered in an action under the Fatal Accidents Act, if such expenses have been incurred by the parties for whose benefit the action is brought.[41]

The basis of any claim for damages under the Act is pecuniary loss resulting to the beneficiaries for whose benefit the action is brought as a consequence of the death of the employee.[42] Such loss must be founded upon:

(a) the relevant family relationship between the beneficiary and the deceased[43]:
(b) the pecuniary benefit the deceased might have been expected to have provided for the beneficiary concerned if he had not been killed.[44]

Consequently, the age of the deceased, the amount he was earning, the extent of his own personal expenses,[45] the possibility that he might have married, thereby ceasing to support his parents (if

---

[38] s.3, as substituted by s.3 of the Administration of Justice Act 1982 (c. 53).
[39] Ibid.
[40] Fatal Accidents Act 1976 (c. 30), s.4, as substituted by s.3 of the Administration of Justice Act 1982 (c. 53).
[41] Fatal Accidents Act 1976, s.3, as substituted by s.3 of the Administration of Justice Act 1982 (c. 53). See *Stanton* v. *Ewart F. Youlden Ltd.* [1960] 1 All E.R. 429.
[42] *Pym* v. *Great Northern Ry.* (1862) 2 B. & S. 759.
[43] *Burgess* v. *Florence Nightingale Hospital for Gentlewomen* [1955] 1 Q.B. 349.
[44] *i.e.* what is reasonably to be expected: *Taff Vale Ry.* v. *Jenkins* [1913] A.C. 1: *Berry* v. *Humm & Co.* [1915] 1 K.B. 627; *Baker* v. *Dalgleish SS Co.* [1922] 1 K.B. 361. *Cf. Mehmet* v. *Perry* [1977] 2 All E.R. 529 (damages for loss of wife and mother).
[45] These are not limited to purely personal expenditure but must include all the usual costs associated with an individual's particular life-style—*Nutbrown* v. *Rosier, The Times,* March 1, 1982; *cf. Clay* v. *Pooler* [1982] 3 All E.R. 570; *Harris* v. *Empress Motors Ltd.*: *Cole* v. *Crown Poultry Packers Ltd.* [1983] 3 All E.R. 226, C.A.; *Adsett* v. *West* [1983] 3 W.L.R. 437; [1983] 2 All E.R. 985; *Warwick* v. *Jeffrey, The Times,* June 21, 1983.

they be the beneficiaries concerned),[46] are all relevant factors.[47] In *Cookson* v. *Knowles*,[48] the Court of Appeal said that the "time had come to divide the awards in fatal accident cases into two parts. First, the pecuniary loss up to the date of trial on which interest should run at half-rate (like special damages in personal injury cases); second, the pecuniary loss from the date of trial onwards of which no interest should be payable."[49]

Where the claim is made on behalf of the deceased's widow, the amount which he is regarded as being likely to provide out of his earnings, after deductions have been made for taxation, etc., and personal expenses, is multiplied by a factor to take into account the number of years for which he might have been expected to make such provision. This is an extremely fluid figure.[50] The multiplier has to be selected as that applicable at the date of death and not taken as at the date of trial.[51] A multiplier of 18 in the case of a breadwinner aged 20–30 *might* not be excessive but it was excessive in the case of a breadwinner aged 41 at his death.[52] The court is required by the Fatal Accidents Act 1976, s.3(3)[53] to ignore the widow's remarriage or prospects of remarriage. It has also been held that the fact that she may obtain a degree of financial independence at some point in the future should also be ignored, because the release of her earning capacity cannot be regarded as a gain to her resulting from the death of the deceased.[54] Where the claim is made on behalf of the deceased's widower, he may be able

---

[46] *Dolbey* v. *Godwin* [1955] 1 W.L.R. 553.
[47] So, too, are any independent means of the deceased: *Shiels* v. *Cruickshank* [1953] 1 All E.R. 874. On the effect of the deceased's savings from his wages in calculating the extent of his widow's dependency, see *Gavin* v. *Wilmot Breeden Ltd.* [1973] 1 W.L.R. 1107; [1973] 3 All E.R. 935, C.A.
[48] [1977] 2 All E.R. 820, C.A.; affirmed by the House of Lords [1978] 2 W.L.R. 978; [1978] 2 All E.R. 604, but see also *Pickett* v. *British Rail Engineering Ltd.* [1980] A.C. 136; [1979] 1 All E.R. 774, H.L., where there was upheld an award of interest on general damages for pain, suffering, and loss of amenity, as part of a claim on behalf of the deceased's estate.
[49] *Ibid.*
[50] Depending on the health of the deceased at the time of the accident, but not the character of his work: *Roughead* v. *Railway Executive* (1949) 65 T.L.R. 435; *Bishop* v. *Cunard White Star Co., Ltd.* [1950] P. 240 at p. 248, *per* Hodson J. For the relevance of insurance premiums paid and purchases made, see *Lolley* v. *Keylock* (1984) 81 L.S.Gaz 1518, C.A.
[51] *Graham* v. *Dodd* [1983] 1 W.L.R. 808; [1983] 2 All E.R. 953, H.L.
[52] *Ibid.*
[53] As substituted by s.3 of the Administration of Justice Act 1982, (c. 53). This does not require the effects of the widow's remarriage to be ignored for the purposes of a claim on behalf of the deceased's children: *Thompson* v. *Price* [1973] Q.B. 838.
[54] *Howitt* v. *Heads* [1973] Q.B. 64. *Cf.* where the wife worked before husband's death, assisted by husband—*Cookson* v. *Knowles* [1977] 2 All E.R. 820, C.A.

to claim the expenses of a housekeeper if actually employed but, if the widower leaves work and stops at home to look after the young children of the marriage, he can only claim his loss of wages, even though they are lower than those of a housekeeper.[55] If the claim is by an ummarried 'common-law wife or husband,' "there shall be taken into account (together with any other matter that appears to the court to be relevant to the action) the fact that the dependant had no enforceable right to financial support by the deceased as the result of their living together."[56]

If the beneficiaries are the children of the deceased, the relevant factors are the cost of keeping the children in food, clothes, education, etc., until they can be self-supporting. Remarriage by the mother might affect the damages to be awarded under the Acts where the step-father becomes legally obliged to maintain and support the children of the deceased: though not where such obligation is moral, not legally enforceable: *Reincke* v. *Gray*.[57] The death of the wife after that of the deceased employee in respect of whose death the action is brought under the Act may affect the question of damages, by reducing what the wife may obtain under the Act, and what may be obtained by the children in view of their accelerated receipt of the damages payable to the widow because of her own death.[58]

Before amendment of the 1976 Act by the Administration of Justice Act 1982,[59] no damages were recoverable for mental distress at the loss of a relative.[60-61] The 1982 Act has now introduced a claim for a fixed sum (currently £3,500) for "damages for bereavement." This sum is recoverable in addition to any other damages under the 1976 Act[62] but only by the spouse of the deceased or the parent(s)[63] of a minor child who was never married.[64]

---

[55] *Bailey* v. *Barking and Havering Area Health Authority*, The Times, July 22, 1978.
[56] Fatal Accidents Act 1976 (c. 30), s.3(4), as substituted by s.3 of the Administration of Justice Act 1982 (c. 53).
[57] [1964] 2 All E.R. 687, distinguishing *Mead* v. *Clark, Chapman & Co., Ltd.* [1956] 1 W.L.R. 76. Voluntary contributions by an uncle had to be ignored in *Rawlinson* v. *Babcock & Wilcox Ltd.* [1967] 1 W.L.R. 481.
[58] *Voller* v. *Dairy Produce Packers Ltd.* [1962] 3 All E.R. 938.
[59] s.3(1), introducing a new s.1A into the Fatal Accidents Act 1976.
[60-61] *Blake* v. *Midland Ry.* (1852) 18 Q.B. 93. *Cf.* the claim in Scottish law for loss of society.
[62] But the claim does not survive the death of the claimant under the 1976 Act, *e.g.* of a wife dying after her deceased husband—1982 Act, s.4(1).
[63] Both parents if the child was legitimate (the sum of £3,500 to be divided equally between them); the mother if the child was illegitimate. (1982 Act, s.3(1)—new s.1A(2) and (4) of the 1976 Act.)
[64] See n.59.

(4) INTERRELATION OF STATUTORY RIGHTS

(a) Distinctions

Under the Fatal Accidents Act, damages are recoverable for the benefit of dependants, not the estate of the deceased, *i.e.*, his successors in title. Nor does it matter if the death of the deceased was the result of suicide, as long as the suicide was caused by the wrongful act or neglect of the employer.[65] Such damages are calculated on the loss to the dependants resulting from the death of the workman. Under the Law Reform (Miscellaneous Provisions) Act 1934 damages are recoverable by the estate of the deceased, are part of that estate, and are based upon what the deceased himself suffered by virtue of his injury and accelerated death. Only in respect of funeral expenses are the rights conferred by these different Acts exactly the same.

(b) Independence

The rights conferred by the Law Reform (Miscellaneous Provisions) Act 1934 are in addition to and not in derogation of rights conferred on dependants of deceased employees by the Fatal Accidents Act.[66] Before the Administration of Justice Act 1982,[67] if a dependant inherited damages payable to the deceased's estate under the 1934 Act, those damages had to be deducted from the dependant's damages under the Fatal Accidents Act 1976.[68] Now, under the new section 4 of the 1976 Act introduced by the 1982 Act,[69] no benefit accruing from the deceased's estate has to be deducted, with the result that a dependant may recover damages under both the 1934 and the 1976 Acts. However, damages under the 1934 Act are much reduced in scope by the 1982 Act, *e.g.* no damages are payable for loss of expectation of life or for earnings for the "lost" years, so there will be no real duplication of damages.

---

[65] *Pigney* v. *Pointer's Transport Services, Ltd.* [1957] 1 W.L.R. 1121.
[66] Law Reform (Miscellaneous Provisions) Act, 1934, s.1(5).
[67] 1982, c. 53.
[68] *Davies* v. *Powell Duffryn Collieries Ltd.* [1942] A.C. 601, H.L.
[69] 1982 Act, s.3.

6. Claims by Employees in Industrial Tribunals

(a) Introductory

Industrial tribunals consisting of a legally qualified chairman and two lay "members" were first set up by the Industrial Training Act 1964 (now replaced by the Industrial Training Act 1982) to deal with appeals against industrial training levies which the various training boards are authorised to impose. This jurisdiction still exists but the industrial tribunals have by various statutory provisions been given increased jurisdictions some of which are relevant to health and safety at work. We have already seen that industrial tribunals have jurisdiction to award payment of remuneration due to employees who have been medically suspended under the terms of certain Health and Safety Regulations and Codes of Practice.[1] There are two specific heads of claims by employees in industrial tribunals. The first is the claim for unfair dismissal and below[2] are discussed those cases where a dismissal has an element of health and safety involved in it. The second is the separate jurisdiction given to an industrial tribunal to adjudicate on rights for time off, etc., for safety representatives.[3] In addition, industrial tribunals can hear appeals by employers against improvement and prohibition notices issued by officers of the Health and Safety Executive.[4]

(b) Unfair dismissal

A general right to claim compensation (or a recommendation for reinstatement or re-engagement) before an industrial tribunal for

---

[1] See Chap. 2.
[2] pp 97–98.
[3] pp. 98–100 below.
[4] See Chap. 8.

an "unfair" dismissal is given by Part V of the Employment Protection (Consolidation) Act 1978 (ss.54 to 80 inclusive). A dismissal consists of an actual dismissal or in certain circumstances a non-renewal of a fixed term. Also included is a situation where "the employee terminates . . . [the contract of employment], with or without notice, in circumstances such that he is entitled to terminate it without notice by reason of the employer's conduct" (constructive dismissal).[5] The question of whether or not a dismissal is fair is dealt with in detail in section 57 of the 1978 Act and requires that an employer must justify the dismissal by assigning a reason to it which is related either to the capability or qualifications of the employee to perform work of the kind he was employed to do or to the employee's conduct or to the fact that the employee was redundant or (which may well be relevant in a health and safety context) where the reason "was that the employee could not continue to work in the position which he held without contravention (either on his part or on that of his employer) of a duty or restriction imposed by or under an enactment."[6] This normally presumably applies to, *e.g.* the dismissal of a lorry driver who loses his driving licence but it could obviously apply to a situation where the particular type of employee could not be lawfully employed, *e.g.* young persons or women in certain occupations.[7]

An employee who is *suspended* by reason of a requirement of certain health and safety regulations may be entitled to remuneration during suspension. Moreover if the employer dismisses instead of suspends the qualifying period of service for an unfair dismissal claim by such an employee is reduced from the normal period of two years to a period of one month.[8] It must be remembered that, even if the employer shows a reason under section 57 of the 1987 Act, he still has to show then that having regard to that reason the dismissal was fair, a question which "shall depend on whether in the circumstances (including the size and administrative resources of the employer's undertaking) the employer acted reasonably or unreasonably in treating it as a sufficient reason for dismissing the employee; and that question shall be determined in accordance with equity and the substantial merits of the case."[9]

---

[5] See p. 98.
[6] 1978 Act, s.57(2)(*d*).
[7] See, *e.g.* s.74 of the Factories Act 1961—prohibition of employment of women and young persons in certain processes connected with lead manufacture and compare the Sex Discrimination Act 1986.
[8] 1978 Act, s.64(2) and see pp. 58–60 above.
[9] 1978 Act, s.57(3), as amended.

(c) Constructive dismissal

A "constructive" dismissal can occur where it is the employee who terminates the contract of employment but he does so with or without notice, in circumstances such that he is entitled to terminate it without notice by reason of the employer's conduct.[10] The employer's conduct which justifies the employee in terminating his contract without notice and enables him to make a claim for constructive dismissal has been the subject of voluminous case law. In short the position is that there must be a serious or fundamental breach by the employer of an express or implied term of the employment contract to bring this rule into play. If that has occurred it will then of course be very difficult for the employer to show that nevertheless the constructive dismissal was fair under the general terms of section 57 of the 1978 Act.[11] In practice, if an employee in a health and safety context can show that the employer's conduct in a health and safety matter was such as to justify the employee leaving without notice then success in an unfair dismissal claim (subject of course to completion of the qualifying period of service, etc.) must be virtually automatic. A failure by an employer for example to take safety precautions to protect his employees against robberies may on the facts constitute a constructive dismissal.[12] Similarly, neglect by the employer to observe the duty of care of an employer to take precautions for his employee's safety or breach of health and safety Acts or regulations, if sufficiently serious, would doubtless entitle an employee to leave and then claim compensation for unfair dismissal. Presumably such an employee would have to show that he had first asked the employer to put matters right and that within a reasonable time the employer had not done so.

(d) Rights of safety representatives

The Safety Representatives and Safety Committees Regulations 1977[13] provide that trade unions may appoint safety representatives from amongst the employees of an employer who recognises the trade union concerned. Those safety representatives, when appointed, have a general right to be consulted[14] and in addition

---

[10] 1978 Act, s.55(2)(c).
[11] See above.
[12] Compare *Keys* v. *Shoe Fayre Ltd.* [1978] I.R.L.R. 476 (Industrial Tribunal) distinguished in *Dutton and Clark* v. *Daly* [1985] I.C.R. 780, E.A.T.
[13] S.I. 1977 No. 500, made under s.2(4) of the Health and Safety at Work etc. Act 1974.
[14] Health and Safety at Work etc. Act 1974, s.2(6).

are given detailed functions under the 1977 Regulations.[15] It is provided[16] that:

> "4. (2) An employer shall permit a safety representative to take such time off with pay during the employee's working hours as shall be necessary for the purposes of—
> (a) performing his functions under section 2(4) of the [Health and Safety at Work etc. Act 1974] and [the detailed functions under the 1977 regulations];
> (b) undergoing such training in respects of those functions as may be reasonable in all the circumstances having regard to any relevant provisions of a Code of Practice relating to time off for training approved for the time being by the Health and Safety Commission under section 16 of the 1974 Act."[17]

It is further provided[18] that a safety representative may present a complaint to an industrial tribunal either that the employer has failed to permit him to take time off as he was entitled to do under the Regulations or alternatively, even if he was given the time off, that the employer has failed to pay him for that time off. Such a complaint must be made to the industrial tribunal within three months of the date when the failure occurred or within such further period as the tribunal considers reasonable in a case where the tribunal is satisfied that it was not reasonably practicable for the complaint to be presented within a period of three months. If the tribunal finds that the employer did fail to permit the employee to take time off, then the industrial tribunal *must* make a declaration to that effect and *may* make an award of compensation to be paid by the employer to the employee of such sum "as the tribunal considers just and equitable in all the circumstances having regard to the employer's default in failing to permit time off to be taken by the employee and to any loss sustained by the employee which is attributable to the matters complained of."[19] If the employer did allow the employee to take time off (or the employee just took the time off) and the employer failed to pay the employee for that time, then if the industrial tribunal finds that the employer has

---

[15] *e.g.* under reg. 4, to investigate potential hazards and complaints, and to make representations to the employer, to attend meetings of safety committees and under reg. 5 to make inspections of the workplace.
[16] By reg. 4(2) of the 1977 Regulations.
[17] The question of what is "pay" for this purpose is to be determined by the Schedule to the 1977 Regulations.
[18] By reg. 11 of the 1977 Regulations.
[19] 1977 Regulations, reg. 11(3).

failed to pay for that period then the tribunal *must* order the employer to pay to the employee the relevant amount (see Schedule to the 1977 Regulations). Disputes may of course arise as to whether or not the time off claimed or taken was truly for a purpose within the 1974 Act and the 1977 Regulations and in one case[20] the Employment Appeal Tribunal had in deciding such a claim to adjudicate on the comparative merits of "in-company" health and safety courses versus such courses provided by a technical college.

---

[20] *White* v. *Pressed Steel Fisher Ltd.* [1980] I.R.L.R. 176, E.A.T.

7. Claims by Employees, etc., for Social Security Benefits

(a) Introductory

Employees in the narrow sense of that word, *i.e.* those who are in the old-fashioned terminology servants of a master, may, if they suffer an accident arising out of and in the course of their employment[1] or having worked for a specific time in a prescribed employment contract a prescribed industrial disease,[2] be able to claim enhanced social security benefits.[3] Moreover, in certain circumstances[4] the State will pay a lump sum compensation to those employees who would have had a right of action in court against an employer for, *e.g.* negligence when the employee has caught certain prescribed lung diseases but the employee is unable to enforce any such claim against the employer because the employer has ceased to carry on business.[5] Enhanced social security benefits where an industrial factor is involved are only available to employees in the true sense and not to self-employed persons.[6] An industrial factor is present where it can be shown that there was either an accident arising out of or in the course of employment or that the employee contracted a prescribed disease in a prescribed occupation. There is no contribution test for these benefits. If the claimant is not an employee in the true sense or, even if he is, he cannot show either an accident arising out of and in the course of employment or a prescribed disease, then the claimant will be able to obtain only sickness and invalidity benefits.[7] These benefits are paid at a lower rate and depend on complete incapacity for any

---

[1] See pp. 103–104.
[2] See p. 103.
[3] See pp. 104–105.
[4] See pp. 105–107.
[5] The employee must also be or have been entitled to disablement benefit—for the whole subject see the Pneumoconiosis, etc., (Workers' Compensation) Act 1979.
[6] Social Security Act 1975, ss.76–78 and compare s.156.

kind of work which the employee could reasonably be expected to do.[7-8] They are normally subject to a contributions test.[9]

An employee is, however, able to claim social security benefits in addition to any claim for damages or other compensation he may have against his employer for injuries or diseases contracted at work. But by section 2 of the Law Reform (Personal Injuries) Act 1948[10] one-half of industrial injury and disablement benefits (except constant attendance allowance) and of sickness and invalidity benefits received or probably to be received for the first five years after the injury must be deducted from damages for loss of earnings or profits.[11] Under the ordinary principles of assessment of damages it has been held that all of employment benefit, statutory sick pay, industrial rehabilitation allowance and supplementary benefit are deductible from damages for loss of earnings.[12]

We will now go on to consider the benefits themselves and then the special system of adjudication for social security benefits.

(1) DISABLEMENT BENEFIT

From the inception of the post-war Social Security Legislation on July 5, 1948, there has always been a specialised system of social security benefits where there is a recognised industrial connection, *i.e.* connected specifically with injuries or prescribed diseases sustained or contracted at work. This system replaced the pre-war system of rights to limited compensation from employers under the Workmen's Compensation Acts. The same test was employed under those Acts as is now used by the social security legislation,[13] namely that the accident must "arise out of and in the course of the employment" with the result that many pre-war decisions of the courts (including those of the House of Lords) on the meaning of that phrase are still relevant to Social Security Law today.

The present law is to be found in Chapters IV and V of the

[7-8] Social Security Act 1975, s.17(1).
[9] Though contributions are not needed for qualification for Severe Disablement Allowance—Social Security Act 1975, s.36.
[10] See p. 49.
[11] But no credit need be given for such benefits after the five years' period—*Barnes* v. *Bromley London Borough Council*, *The Times*, November 19, 1983; *Denman* v. *Essex Area Health Authority*: *Haste* v. *Sandell Perkins Ltd.* [1984] 3 W.L.R. 73; *Jackman* v. *Corbett* [1987] 2 All E.R. 699, C.A.
[12] But not attendance or mobility allowances as they are for specific services—*Bowker* v. *Rose*, *The Times*, February 3, 1978, C.A. See further p. 49.
[13] See now ss.50–55 of the Social Security Act 1975.

Social Security Act 1975[14] and in the Social Security (Industrial Injuries) (Prescribed Diseases) Regulations 1985.[15] The 1975 Act and the 1985 Regulations have been extensively amended and in particular by the Social Security Act 1986.

To be able to claim disablement benefit the employee[16] must have suffered either "accident arising out of and in the course of employment,"[17] or must be able to show that he has suffered a prescribed disease or injury. That occurs where the disease itself is by definition prescribed in the 1985 Regulations and he has worked in an occupation also described in detail in those Regulations.[18] If an employee has suffered injury or disease at work which is not attributable to an accident arising out of and in the course of his employment and is the result of a "process" rather than an "accident" then no claim for social security disablement benefit can be maintained unless the employee can show that he has suffered a prescribed disease or injury. For details of these diseases and injuries the 1985 Regulations themselves must be consulted but Schedule 1 to the Regulations prescribes some 60 diseases or injuries with corresponding prescribed occupations being sub-classified into "conditions due to physical agents," "conditions due to biological agents," "miscellaneous conditions" and "pneumoconiosis, byssinosis and allied diseases." It suffices that an employee has worked in a prescribed occupation for no matter how short a period of time except that he must have worked in the prescribed occupation for not less than 10 years if it is occupational deafness that is in issue. Nor will he be able to pray in aid the presumption that the disease was due to the nature of his employed earner's employment unless he has been employed in the employment for at least one month.[19] What constitutes an accident and when it arises out of and in the course of the employment has been the subject of voluminous case law, consisting of decisions of the courts under the Workmen's Compensation legislation and decisions of the Social Security Commissioners (formerly National Insurance Commissioners). It is not possible in this book to deal in detail with these decisions and reference should be made to, *e.g.* the specialised annotations in Bonner, etc., *Non-Means Tested*

---

[14] ss.50–78 and *cf.* s.156.
[15] S.I. 1985 No. 967.
[16] For special cases of "employment" see the Social Security (Employed Earner's Employments for Industrial Injuries Purposes) Regulations 1975, S.I. 1975 No. 467.
[17] Social Security Act 1975, s.50(1) and see also ss.51–55.
[18] See regs. 2 and 4 of the 1985 Regulations.
[19] See reg. 2(*c*) and 4(1) of the 1985 Regulations.

Benefits: The Legislation (2nd ed., 1987)[19a] and to the series of reported Commissioners' decisions published by H.M. Stationery Office. Suffice to say that "accident" is widely interpreted and can include not only accidents in the ordinary sense but also deliberate acts.[20] It can also include unforeseen but ascertainable internal psychological changes such as a heart attack or a sudden strain of a muscle but not adverse conditions of the body which come on gradually by way of a process rather than as a result of one or more specific incidents.[20a] Whether an admitted accident arises "in the course of" employment may also give rise to problems.[20b]

If entitlement to disablement benefit is shown by proof of an industrial accident or contracting a prescribed disease, then the claimant is entitled to disablement benefit according to the detailed rules set out in Chapter IV of the Social Security Act 1975. Again for the details of this the reader must be referred to specialised works. Until the changes introduced by the Social Security Act 1982 so long as the employee was wholly incapacitated for work he would be able to obtain for a maximum period of six months industrial injury benefit of a weekly amount at a higher rate than that of sickness or invalidity benefit but that differential was abolished by the 1982 Act. Now during incapacity for work, even though as a result of an industrial accident or a prescribed disease, only sickness or invalidity benefits are obtainable.[21] If after the initial six months' period the claimant is still suffering from a loss of faculty (its amount to be assessed at a percentage by the medical adjudicating authorities)[22] then he is entitled to disablement benefit in accordance with scales prescribed by the Social Security Act 1975 (as amended by the Social Security Act 1986) and by the Social Security Benefit (General Benefit) Regulations 1982.[23] Under the changes introduced by the Social Security Act 1986[24] no disablement benefit is payable where the assessment of disablement is less than 14 per cent. For assessments of disablement between 14 per cent. and 100 per cent. a weekly disablement

---

[19a] pp. 78–84.
[20] *Trim Joint District School Board* v. *Kelly* [1914] A.C. 667, H.L. (Schoolmaster put to death by assault by pupils).
[20a] See, *e.g.* reported Commissioners' Decisions R(I) 43/55 and R(I) 6/82.
[20b] See, *e.g. R.* v. *Industrial Injuries Commissioner, ex p.* A.E.U. (No. 2) [1966] Q.B. 31, C.A. (smoking break) and *Nancollas* v. *Insurance Officer* [1985] 2 All E.R. 833, C.A. (travelling accident *cf.* Social Security Act 1975, s.53).
[21] Though it is still important to show the industrial connection in the case of married women whose contributions are insufficient to entitle them to contributory benefits—see Social Security Act 1975, s.50A.
[22] Adjudicating medical practitioner with appeal to medical appeal tribunal.
[23] S.I. 1982 No. 1408.
[24] See Sched. 3 of that Act.

pension is payable its amount varying with the degree of disablement. There are detailed provisions in the regulations[25] as to the percentage extent of disablement for specific injuries, *e.g.* loss of a leg or an eye and rules as to "paired" organs. Currently for example the rate of disablement benefit for 14–20 per cent. disablement is £12.90 per week and for 100 per cent. disablement is £64.50 per week (less in both cases if claimant under 18).

(2) SICKNESS AND INVALIDITY BENEFITS

An employee who is injured or who contracts a disease at work, if he cannot bring himself within the special provisions for receipt of industrial disablement benefit (above), will be nevertheless able to claim sickness and invalidity benefits (sickness benefit for the first 192 days, thereafter invalidity benefit at an enhanced rate) for such time as he is incapable of work by reason of specific disease or bodily or mental disablement.[26] But entitlement to these benefits is dependent on satisfaction of contribution conditions.[27] If there is an insufficient contribution record then the claimant will have to fall back on being able to show 80 per cent. disablement, when he will be entitled[28] to Severe Disablement Allowance of an amount similar to Invalidity Benefit.

(3) PAYMENTS BY THE DEPARTMENT OF EMPLOYMENT UNDER THE PNEUMOCONIOSIS, ETC., (WORKERS' COMPENSATION) ACT 1979

The Pneumoconiosis, etc., (Workers' Compensation) Act 1979 introduced a new scheme to ensure proper compensation for those who had contracted certain lung diseases at work and who might well have been able to claim damages from an employer in respect of the contracting of this disease, *e.g.* because of negligence or an unsafe system of work, but "every relevant employer of [the employee] has ceased to carry on business."[29] "Cease to carry on

---

[25] Social Security (General Benefit) Regulations 1982, S.I. 1982 No. 1408.
[26] Social Security Act 1975, s.17(1)—"work" means "work which the claimant can reasonably be expected to do"—proviso to s.17(1).
[27] See Social Security Act 1975, ss.1–13 and numerous regulations made under the 1975 Act relating to contributions and credits.
[28] Under the new provisions of s.36 of the Social Security 1975 introduced by s.11 of the Health and Social Security Act 1984.
[29] 1979 Act, s.2(1)(*b*).

business" is not defined in the 1979 Act or elsewhere. The expression presumably extends to cases where the employer's business activities are defunct (at least in relation to the occupation where the disease was contracted), even though the employer, if a company, has not gone into liquidation or been struck off the Register of Companies. In addition the employee has to show that he has not brought any action in court, relating to the disease, against an employer,[30] nor must he have compromised (*i.e.* settled for compensation) any claim against the employer. Additionally the employee must show that he is entitled to or has been entitled to social security disablement benefit[31] whether or not he is actually being paid that benefit for disablement[32] for one of the relevant diseases (see below) contracted in an occupation for which it is prescribed by the Social Security (Industrial injuries) (Prescribed Diseases) Regulations 1985.[33] The relevant diseases[34] are as follows:

(a) Pneumoconiosis (including silicosis, asbestosis, and kaolinosis).
(b) Byssinosis (caused by cotton or flax dust).
(c) Diffuse mesothelioma.
(d) Primary carcinoma of the lung where there is accompanying evidence of one or both of asbestosis and bilateral diffuse pleural thickening.
(e) Bilateral diffuse pleural thickening.

Provisions are also made for lump sum payments to be made by the Department of Employment to dependants where the disabled person has died.[35]

The amounts of the lump sums are prescribed currently by the Pneumoconiosis, etc., (Workers' Compensation) (Payment of Claims) Regulations 1988[36] and reference must be made to the detailed and sub-categorised Schedule to those Regulations. Examples, however, are that there would be a payment of £35,879 to a person who was found to be 100 per cent. dis-

---

[30] For this purpose an action which has been dismissed otherwise than on the merits (as for example for want of prosecution or under any enactment relating to the limitation of actions) shall be disregarded—1979 Act, s.2(4).
[31] See pp. 103–104.
[32] Social Security Act 1975, s.57.
[33] See pp. 103–104.
[34] Specified in the 1979 Act, s.1(3), as added to by the Pneumoconiosis, etc., (Workers' Compensation) (Specified Diseases) Order 1985 S.I. 1985 No. 2034.
[35] See 1979 Act, s.3 for definition of "dependant," being basically spouse if there was one, otherwise children or if no children, then relatives provided either young or incapable of self-support.
[36] S.I. 1988 No. 668.

abled from one of the above diseases and was aged 37 and under (the amount of the payment diminishes if the claimant is older) to a payment of £1263 to a person found to be 10 per cent. or less disabled by one of the diseases and aged 77 or over. If it is a claim by a dependant where there has been a death of a disabled person then again the scale varies according to the extent of the deceased's disablement and his age at his last birthday preceding death. The younger he was the greater the amount of the payment to the dependants.

(4) SYSTEM OF ADJUDICATION

Claims to social security benefits and questions arising out of those claims have to be determined by the specialised system of adjudication laid down by the Social Security Act 1975; the Social Security (Adjudication) Regulations 1986[37] and the Social Security Commissioners Procedure Rules 1987.[38] These provide for a specialised system of adjudication and it is only after an appeal to a Social Security Commissioner has been determined that recourse lies by way of appeal in the normal way to the ordinary court system, appeal lying from a Social Security Commissioner to the Court of Appeal and thence to the House of Lords (in both cases with leave only and on a question of law). The tiers of adjudication thus provided for are as follows. Initially any claim or question relating to a social security benefit must be referred to a local adjudication officer[39] who must within a reasonable time and so far as it is practicable within 14 days either decide the question or refer it to a social security appeal tribunal. If the local adjudication officer has decided the point himself, the claimant may appeal against the adjudication officer's decision to a social security appeal tribunal.[40] Such a tribunal is independent of the Departments and consists of a legally qualified chairman sitting with two members selected from panels constituted by the President of Social Security Appeal Tribunals. The panels are composed of persons appearing to the President to have knowledge or experience of conditions in the area and to be representative of persons living or working in the area. Where practicable at least one of the members of the tribunal hearing a case must be of the same sex as

---

[37] S.I. 1986 No. 2218.
[38] S.I. 1987 No. 214.
[39] Social Security Act 1975, ss.97 and 98.
[40] See s.100 of and Sched. 10 to the Social Security Act 1975.

the claimant.[41] The social security appeal tribunal's decision can be by a majority or unanimous. If, with the consent of the claimant, the tribunal consists of the chairman and only one member, then the chairman has a casting vote.[42] The decision will deal with both factual and legal issues but the appeal to the Social Security Commissioners lies only on a question of law.[43–44] Moreover, leave either of the social security appeal tribunal chairman of, if he refuses it, of a Social Security Commissioner is necessary for such an appeal. The Social Security Commissioners are legally qualified 'judges' independent of the Departments and appointed by the Queen.[45] They normally give decisions on appeal singly but it is possible for the Chief Commissioner to appoint a Tribunal of three Commissioners to hear questions of law of exceptional difficulty.[46] Although appeal to the Commissioners lies only on a question of law the Commissioners have power where they consider it expedient to make their own findings of fact and to give decisions on factual as well as legal issues where once the necessary precondition has been established of the social security appeal tribunal having erred in law. From the Social Security Commissioners appeal lies with leave either of a Commissioner or of the Court of Appeal on a point of law to the Court of Appeal and thence with similar leave to the House of Lords. A number of important questions relating to social security benefits have been decided by the House of Lords.[47]

Legal aid for legal representation before the social security adjudicating authorities is not available but advice is obtainable under the general provisions of the legal aid scheme. If the case proceeds to the Court of Appeal or House of Lords legal aid for representation is obtainable in the normal way. Frequently parties before the social security appeal tribunals and the Social Security Commissioners are represented other than by lawyers and by those who specialise in the field, *e.g.* the Citizens Advice Bureau or the Child Poverty Action Group.

Adjudication on medical questions such as the extent of disablement or whether or not there has been a recrudescence of an

---

[41] Social Security Act 1975, Sched. 10, para. 1(8).
[42] Reg. 24(2) of the Social Security (Adjudication) Regulations 1986, S.I. 1986 No. 2218.
[43–44] Social Security Act 1975, s.101 as amended by para. 7 of Sched. 5 to the Social Security Act 1986.
[45] Social Security Act 1975, s.97(3).
[46] Social Security Act 1975, s.116.
[47] See for example *Presho* v. *Insurance Officer* [1984] A.C. 310, H.L. and *Lees* v. *Secretary of State for Social Services* [1985] A.C. 930, H.L.

industrial disease are not of course adjudicated upon within the normal system of social security adjudication since those who adjudicate have no medical qualifications. Such issues are the subject of a special system of medical adjudication and now[48] are confided to adjudicating medical authorities, *i.e.* normally one medical practitioner though in certain cases[49] adjudication has to be by two such practitioners. Appeal lies from an adjudicating medical authority to a medical appeal tribunal, comprised of a legally qualified chairman and two qualified medical members. Appeal from a medical appeal tribunal lies but on a question of law only to a Social Security Commissioner.[50] The system of medical adjudication applies to disablement benefit, but for certain other types of benefit, *e.g.* attendance allowance and mobility allowance, there are special provisions for medical adjudication.

Adjudication on questions arising under the Pneumoconiosis, etc., (Workers' Compensation) Act 1979 is governed by sections 4 and 5 of the Act and by the Pneumoconiosis, etc., (Workers' Compensation) (Determination of Claims) Regulations 1985.[51] That legislation provides for the holding of an enquiry by a person appointed by the Secretary of State.[52] There is a 12 months' limit for claims unless the Secretary of State grants an extension, the 12 months running normally from the earliest date on which disablement benefit would have become payable to the claimant where the claimant is alive or within 12 months of the date of death in the case of a claim by a dependant of a deceased claimant. Presumably the medical questions affecting disablement benefit (entitlement to which is a precondition of a payment under the 1979 Act) will still of course be determined by the Social Security adjudicating authorities as detailed above.

---

[48] Under the provisions of ss.108–111 of the Social Security Act 1975 and regs. 27 to 36 of the Social Security (Adjudication) Regulations 1986, S.I. 1986 No. 2218.
[49] Reg. 28 of the Social Security (Adjudication) Regulations 1986, S.I. 1986 No. 2218.
[50] Social Security Act 1975, s.112.
[51] S.I. 1985 No. 1645.
[52] 1979 Act, s.4(2), from whom an appeal lies to the High Court on a question of law (s.4(3) of the 1979 Act applying s.94 of the Social Security Act 1975).

8. Administration and Enforcement of Safety, Health and Welfare Legislation

(1) INTRODUCTORY

Administration of all the safety, health and welfare legislation (both principal and subordinate) applicable to employers, employees and the self-employed was unified and strengthened by the Health and Safety at Work etc. Act 1974.[1] This Act, which has been described as the most significant statutory advance in the field since Shaftesbury's Factory Act of 1833, not only lays down new statutory duties of general import,[2] but also adopts for the time being all the existing statutory provisions,[3] such as the Mines and Quarries Act 1954, the Factories Act 1961, and the Offices, Shops and Railway Premises Act 1963, together with regulations, orders, etc., made under these Acts, though the intention is eventually to replace the existing statutory provisions by new regulations and codes of practice to be made under the 1974 Act.

The Membership by the United Kingdom of the European Economic Community (EEC) means that future laws in the field of safety, health and welfare at work will have to comply with the harmonisation aims of the Community. The EEC Council in 1978 resolved on an action programme in the Community on health and safety at work, stressing in particular action to be taken by 1982 under the following heads:

(i) Accident and disease aetiology connected with work-research;
(ii) Protection against dangerous substances;

---

[1] 1974, c. 37. This Act was passed to implement the recommendations of the Robens Committee on Safety and Health at Work 1972 Cmnd., 5034.
[2] *Ibid.* ss.2–9 (the self-employed, and members of the public adversely affected by industrial operations, are protected).
[3] For a full list of all the "relevant statutory provisions," see Sched. 1 to the 1974 Act.

(iii) Prevention of the dangers and harmful effects of machines;
(iv) Monitoring and inspection—improvement of human attitudes.

The Resolution also contains, in the Annex to it, a detailed and valuable action programme. Many recent regulations, *e.g.* the Control of Industrial Major Accident Hazards Regulations 1984[3a] and the Ionising Radiations Regulations 1985,[3b] have been made in implementation of EEC Directives.

(2) HEALTH AND SAFETY ORGANISATION

(a) Health and Safety Commission

The Health and Safety at Work etc. Act 1974 set up an entirely new health and safety organisation. There is created a two-tier structure, shown diagrammatically on the following page, headed by the Health and Safety Commission which can give certain directions[4] to its operational arm—the Health and Safety Executive. In addition the Commission has wide statutory duties, *e.g.* to see that the general purposes of the 1974 Act are fulfilled, and to arrange for research, advice, and information.[5] The Commission is subject to directions from the Secretary of State for Employment as to its functions, or those affecting the safety of the State.[6–9]

The Commission also has wide statutory powers, *e.g.* to make agency agreements for government departments or others, to perform functions on its behalf or on behalf of the Health and Safety Executive (*infra*), to take and pay for expert advice,[10] to direct the holding of an investigation or inquiry into accidents or other occurrences,[11] and to approve and issue codes of practice, non-compliance with which may raise a presumption of guilt in criminal proceedings for breach of the statutory duties.[12]

[3a] S.I. 1984 No. 1902
[3b] S.I. 1985 No. 1333.
[4] 1974 Act, s.11(4).
[5] *Ibid.* s.11.
[6–9] *Ibid.* s.12.
[10] *Ibid.* s.13(1)(*d*).
[11] *Ibid.* s.14.
[12] *Ibid.* ss.16–17.

HEALTH AND SAFETY COMMISSION/EXECUTIVE
(Outline Main Organisation Chart)

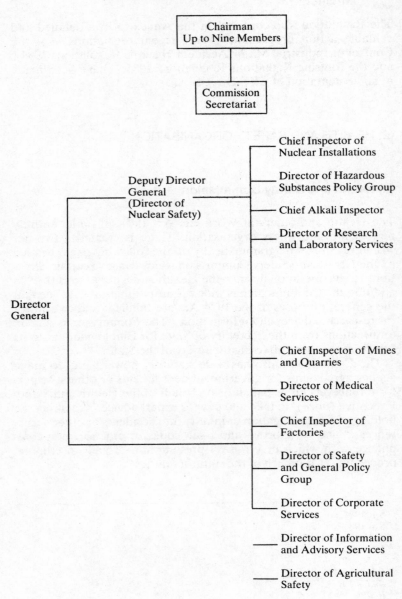

(b) Health and Safety Executive

The Health and Safety Executive, the operational arm of the Health and Safety Commission, is the body primarily charged with the duty of enforcing the relevant statutory provisions.[13] Local authorities are responsible for enforcement of certain classes of activities designated by regulations,[14] *e.g.* offices, catering and hotels, and some shops, while responsibility for fire prevention and precautions is to be shared between fire authorities, the Home Office, and the Health and Safety Executive.[15]

The Health and Safety Executive is empowered to appoint inspectors[16] and the former Inspectorates of Factories, Mines and Quarries, Nuclear Installations, Alkali and Clean Air, and Explosives have been transferred to the Health and Safety Executive.[17] The Executive is also to have a major research, information, education and advisory role, and is to include the Safety in Mines Research Establishment. In addition the Executive is responsible for the Employment Medical Advisory Service set up in 1972.[18]

The Commission is empowered[19] to make agency agreements with government departments or others to perform the Executive's functions on its behalf.

(c) Fire authorities

Normally, the fire authority concerned with the observance of fire precautions in factories, offices, shops and railway premises[20] is, in England and Wales, the County Council,[21] and, in Scotland, the Islands or Regional Council.[22] However, in the case of factory,

---

[13] *e.g.* the Mines and Quarries Act 1954, the Factories Act 1961 and the Offices, Shops and Railway Premises Act 1963 (see full list in Sched. 1 to the 1974 Act), as well as the provisions of the 1974 Act itself.
[14] See the Health and Safety (Enforcing Authority) Regulations 1977, S.I. 1977 No. 746.
[15] *Ibid.* s.78.
[16] *Ibid.* s.19.
[17] Enforcement of the Employers' Liability (Compulsory Insurance) Act 1969 has in pursuance of an agency agreement been transferred from the Department of Employment to the Health and Safety Executive (S.I. 1975 No. 194).
[18] 1974 Act, ss.55–56.
[19] By s.13 of the 1974 Act.
[20] The law is now contained in the Fire Precautions Act 1971 as amended by the Fire Safety and Safety of Places of Sport Act 1987; the Fire Precautions (Factories, Offices, Shops and Railway Premises) Order 1976, S.I. 1976 No. 2009; and the Fire Precautions (Non-Certificated Factory, Office, Shop and Railway Premises) Regulations 1976, S.I. 1976 No. 2010.
[21] But see s.41(1) of the Fire Precautions Act 1971.
[22] *Ibid.*

office, shop and railway premises within a defined list of special hazards, the enforcing and certificate granting authority is the Health and Safety Executive.[23]

(3) ENFORCEMENT POWERS

The Health and Safety at Work etc. Act 1974 contains detailed enforcement powers applicable to all the "relevant statutory provisions,"[24–25] *e.g.* the Factories Act 1961 and the Offices, Shops and Railway Premises Act 1963, and as well as repeating the powers given in those statutory provisions (now largely repealed), gives important new powers such as the issue of improvement and prohibition notices described below. This uniform code of enforcement powers is exercisable by Inspectors appointed by the relevant enforcing authority, be it the Health and Safety Executive, a local authority, or some other body. Not all the powers, however, will be conferred on every inspector, for the 1974 Act[26] requires each inspector to be given (and, if required, to produce) a written authority stating which of the powers he may use. "This is intended to ensure that the more stringent powers are always exercised by experienced inspectors, and only in relation to hazards for control of which these more stringent powers are appropriate."[27]

The details of the enforcement powers are contained in sections 20 to 26 of the 1974 Act and include powers of entry, examination, investigation, to take samples, to render harmless dangerous articles and substances, and to require certain necessary information.[28]

Powers[29] are given to inspectors to serve:

(i) improvement notices requiring the remedy of contraventions of the relevant statutory provisions,[30] and

---

[23] Acting under the Fire Certificates (Special Premises) Regulations 1976, S.I. 1976 No. 2003.

[24–25] See full list in Sched. 1 to the 1974 Act. Consequently, the enforcement powers contained in the Acts themselves have been repealed by statutory instruments made under s.80 of the 1974 Act, except in some cases (*e.g.* under the Factories Act 1961 and the Offices, Shops and Railway Premises Act 1963) to retain the powers for fire authorities.

[26] s.19. As to an inspector's authority to prosecute, see *Campbell* v. *Wallsend Slipway and Engineering Co. Ltd.* [1978] I.C.R. 1015, Div.Ct.

[27] Statement of Health and Safety Commission dated December 19, 1974.

[28] 1974 Act, ss.20 and 25. Wider powers to compel the disclosure of information are given to the Health and Safety Commission by ss.27 and 28 of the 1974 Act.

[29] Under ss.21 to 24 of the 1974 Act.

[30] See note 21, *supra*.

(ii) where there is a risk of serious personal injury, notices prohibiting either immediately or, after a specified period, the carrying on of activities in breach of the relevant statutory provisions.[31]

Appeals can be made against improvement or prohibition notices to an Industrial Tribunal. The bringing of the appeal will have the effect of suspending the notice but only if the Tribunal so directs, in the case of a prohibition notice.[32-33] The Tribunal has power to "modify" the requirements of the notice.[34]

The criminal law is the principal instrument for securing compliance with the safety, health and welfare duties imposed by the Health and Safety at Work etc. Act 1974, and by the relevant statutory provisions.[35] Fines and/or imprisonment are provided for[36] on conviction for offences of non-compliance with these duties, or failure to comply with certain requirements of inspectors.[37] The 1974 Act also contains provisions as to prosecutions,[38] and the power of the court to order the cause of an offence to be remedied, or the forfeiture of explosive articles or substances.[39]

---

[31] *Ibid.*
[32-33] s.24.
[34] This includes modifying a vague notice to refer to the relevant statutory provisions—*Chrysler (U.K.) Ltd.* v. *McCarthy* [1978] I.C.R. 939, D.C. but not adding alleged breaches of additional statutory provisions—*British Airways Board* v. *Henderson* [1979] I.C.R. 257. *Cf. Tesco Stores Ltd.* v. *Edwards* [1977] I.R.L.R. 120 (Tribunal can add factual requirements) and *West Bromwich Building Society* v. *Townsend* [1983] I.C.R. 257 (particularity needed for allegation of breach of general duty under the 1974 Act).
[35] Civil actions for damages for breach of the relevant statutory provisions (though not of the general duties under the 1974 Act) are also a possibility—1974 Act, s.47.
[36] By the 1974 Act, s.33. Provisions as to offences and penalties also continue to exist in the relevant statutory provisions, *e.g.* the Factories Act 1961 and the Offices, Shops and Railway Premises Act 1963.
[37] *e.g.* failure to comply with an improvement or a prohibition notice issued under ss.21 and 22 of the 1974 Act—see s.33.
[38] 1974 Act, s.34 (time-limits); s.35 (venue); s.36 (offences due to another's fault); s.37 (bodies corporate); s.38 (only an inspector to prosecute, unless prosecutor has consent of D.P.P.); s.39 (power of inspector to act as advocate); s.40 (onus of proving safety precaution not practicable cast upon accused); s.41 (evidence).
[39] *Ibid.* s.42.

Appendix 1

Factories Act 1961
(c. 34)

Sect.

1. Cleanliness
2. Overcrowding
3. Temperature
4. Ventilation
5. Lighting
7. Sanitary Conveniences
10A. Medical examinations of persons employed in factories
11. Power to require medical supervision
12. Prime movers
13. Transmission machinery
14. Other machinery
15. Provisions as to unfenced machinery
16. Construction and maintenance of fencing
28. Floors, passages and stairs
29. Safe means of access and safe place of employment
57. Supply of drinking water
58. Washing facilities
59. Accommodation for clothing
60. Sitting Facilities
175. Interpretation of expression "factory"

PART I

HEALTH (GENERAL PROVISIONS)

Cleanliness

1.—(1) Every factory shall be kept in a clean state and free from effluvia arising from any drain, sanitary convenience or nuisance.

(2) Without prejudice to the generality of subsection (1) of this section,—

(a) accumulations of dirt and refuse shall be removed daily by a suitable method from the floors and benches of workrooms, and from the staircases and passages;
(b) the floor of every workroom shall be cleaned at least once every week by washing or, if it is effective and suitable, by sweeping or other method.

(3) Without prejudice to the generality of subsection (1) of this section but subject to subsection (4) thereof, the following provisions shall apply as respects all inside walls and partitions and all ceilings or tops of rooms, and all walls, sides and tops of passages and staircases, that is to say,—

(a) where they have a smooth impervious surface, they shall at least once in every period of fourteen months be washed with hot water and soap or other suitable detergent or cleaned by such other method as may be approved by the inspector for the district;
(b) where they are kept in a prescribed manner or varnished, they shall be repainted in a prescribed manner or revarnished at such intervals of not more than seven years as may be prescribed, and shall at least once in every period of fourteen months be washed with hot water and soap or other suitable detergent or cleaned by such other method as may be approved by the inspector for the district;
(c) in any other case they shall be kept whitewashed or colourwashed and the whitewashing or colourwashing shall be repeated at least once in every period of fourteen months.

(4) Except in a case where the inspector for the district otherwise requires, the provisions of subsection (3) of this section shall not apply to any factory where mechanical power is not used and less than ten persons are employed.

(5) [*Repealed from January* 1, 1975, *by the Factories Act* 1961 *etc.* (*Repeals and Modifications*) *Regulations* 1974, *S.I.* 1974 *No.* 1941.]

Overcrowding

2.—(1) A factory shall not, while work is carried on, be so overcrowded as to cause risk of injury to the health of the persons employed in it.

(2) Without prejudice to the generality of subsection (1) of this section but subject to subsection (3) thereof, the number of persons employed at a time in any workroom shall not be such that

the amount of cubic space allowed for each is less than four hundred cubic feet.[1]

(3) If the chief inspector is satisfied that, owing to the special conditions under which the work is carried on in any workroom in which explosive materials are manufactured or handled, the application of subsection (2) of this section to that workroom would be inappropriate or unnecessary, he may by certificate except the workroom from that subsection subject to any conditions specified in the certificate.

(4) [*Repealed from January* 1, 1975, *by the Factories Act* 1961 *etc.* (*Repeals and Modifications*) *Regulations* 1974, *S.I.* 1974 *No.* 1941.]

(5) In calculating for the purposes of this section the amount of cubic space in any room no space more than fourteen feet[2] from the floor shall be taken into account and, where a room contains a gallery, the gallery shall be treated for the purposes of this section as if it were partitioned off from the remainder of the room and formed a separate room.

(6) Unless the inspector for the district otherwise allows, there shall be posted in the workroom a notice specifying the number of persons who, having regard to the provisions of this section, may be employed in that room.

Temperature

3.—(1) Effective provision shall be made for securing and maintaining a reasonable temperature in each workroom, but no method shall be employed which results in the escape into the air of any workroom of any fume of such a character and to such extent as to be likely to be injurious or offensive to persons employed therein.

(2) In every workroom in which a substantial proportion of the work is done sitting and does not involve serious physical effort a temperature of less than sixty degrees[3] shall not be deemed, after the first hour, to be a reasonable temperature while work is going on, and at least one thermometer shall be provided and maintained in a suitable position in every such workroom.

(3) [*Repealed from January* 1, 1975, *by the Factories Act* 1961 *etc.* (*Repeals and Modifications*) *Regulations* 1974, *S.I.* 1974 *No.* 1941.]

---

[1] 11 cubic metres from 12/8/83.
[2] 4.2 metres from 12/8/83.
[3] 16 degrees Celsius.

Ventilation

4.—(1) Effective and suitable provision shall be made for securing and maintaining by the circulation of fresh air in each workroom the adequate ventilation of the room, and for rendering harmless, so far as practicable, all such fumes, dust and other impurities generated in the course of any process or work carried on in the factory as may be injurious to health.

(2) [*Repealed from January* 1, 1975, *by the Factories Act* 1961 *etc.* (*Repeals and Modifications*) *Regulations* 1974, *S.I.* 1974 *No.* 1941.]

Lighting

5.—(1) Effective provision shall be made for securing and maintaining sufficient and suitable lighting, whether natural or artificial, in every part of a factory in which persons are working or passing.

(2) [*Repealed from January* 1, 1975, *by the Factories Act* 1961 *etc.* (*Repeals and Modifications*) *Regulations* 1974, *S.I.* 1974 *No.* 1941.]

(3) Nothing in the foregoing provisions of this section or in any regulations made thereunder shall be construed as enabling directions to be prescribed or otherwise given as to whether any artificial lighting is to be produced by any particular illuminant.

(4) All glazed windows and skylights used for the lighting of workrooms shall, so far as practicable, be kept clean on both the inner and outer surfaces and free from obstruction; but this subsection shall not affect the whitewashing or shading of windows and skylights for the purpose of mitigating heat or glare.

Drainage of floors

6. Where any process is carried on which renders the floor liable to be wet to such an extent that the wet is capable of being removed by drainage, effective means shall be provided and maintained for draining off the wet.

Sanitary conveniences

7.—(1) Sufficient and suitable sanitary conveniences for the persons employed in the factory shall be provided, maintained and kept clean, and effective provision shall be made for lighting them and, where persons of both sexes are or are intended to be employed (except in the case of factories where the only persons employed are members of the same family dwelling there) the conveniences shall afford proper separate accommodation for persons of each sex.

(2) [*Repealed from January* 1, 1975, *by the Factories Act* 1961 *etc.* (*Repeals and Modifications*) *Regulations* 1974, *S.I.* 1974 *No.* 1941.]

Medical examinations of persons employed in factories

10A.—(1) If an employment medical adviser is of opinion that there ought, on grounds mentioned in subsection (2) below, to be a medical examination of a person or persons employed in a factory, he may serve on the occupier of the factory a written notice stating that he is of that opinion and requiring that the occupier shall permit a medical examination in accordance with this section of the person or persons in question, and the examination shall be permitted accordingly.

(2) The grounds on which a medical examination of a person may be required by an employment medical adviser's notice under subsection (1) above are that (in the adviser's opinion) the person's health has been or is being injured, or it is possible that it has been, is being or will be injured, by reason of the nature of the work he is or has been called upon to do or may (to the adviser's knowledge) be called upon to do; and a notice under that subsection may be given with respect to one or more named persons or to persons of a class or description specified in the notice.

(3) A notice under subsection (1) above shall name the place where the medical examination is to be conducted and, if it is a place other than the factory, the day on which and the time at which it is to be begun; and—

(*a*) every person to whom the notice relates shall be informed, as soon as practicable after service thereof, of the contents thereof and of the fact that he is free to attend for the purpose of submitting to the examination; and

(*b*) if the notice states that the examination is to be conducted at the factory, suitable accommodation thereat shall be provided for the conduct of the examination.

(4) A medical examination conducted in pursuance of a notice under subsection (1) above shall be begun within seven days after the day on which the notice is served, and shall be conducted by, or in accordance with arrangements made by, an employment medical adviser, and take place at a reasonable time during working hours.

(5) An employment medical adviser may, by written notice

served on the occupier of a factory, cancel a notice served on the occupier under subsection (1) above; and a notice which relates to two or more named persons may be cancelled in relation to them all or in relation to any one or more of them.

(6) In this section "medical examination" includes pathological, physiological and radiological tests and similar investigations.

Power to require medical supervision

11.—(1) Where it appears to the Minister—
  (*a*) that in any factory or class or description of factory—
    (i) cases of illness have occurred which he has reason to believe may be due to the nature of a process or other conditions of work; or
    (ii) by reason of changes in any process or in the substances used in any process, or by reason of the introduction of any new process or new substance for use in a process, there may be risk of injury to the health of persons employed in that process; or
    (iii) young persons are or are about to be employed in work which may cause risk of injury to their health; or
  (*b*) that there may be risk of injury to the health of persons employed in a factory—
    (i) from any substance or material brought to the factory to be used or handled therein; or
    (ii) from any change in the conditions of work or other conditions in the factory;

he may make special regulations requiring such reasonable arrangements to be made for the medical supervision (not including medical treatment other than first-aid treatment and medical treatment of a preventive character) of the persons, or any class of the persons, employed at that factory or class or description of factory as may be specified in the regulations.

(2) Where the Minister proposes to exercise his powers under this section in relation to a particular factory and for a limited period, he may exercise those powers by order instead of by special regulations, and any such order shall, subject to subsection (3) of this section, cease to have effect at the expiration of such period not exceeding six months from the date when it comes into operation as may be specified in the order.

(3) The Minister may by subsequent order or orders extend the said period, but if the occupier of the factory by notice in writing to him objects to any such extension, the original order shall cease to have effect as from one month after the service of the notice, with-

out prejudice to the making of special regulations in relation to the factory.

Part II

Safety (General Provisions)

Prime movers

12.—(1) Every flywheel directly connected to any prime mover and every moving part of any prime mover, except such prime movers as are mentioned in subsection (3) of this section, shall be securely fenced, whether the flywheel or prime mover is situated in an engine-house or not.

(2) The head and tail race of every water wheel and of every water turbine shall be securely fenced.

(3) Every part of electric generators, motors and rotary converters, and every flywheel directly connected thereto, shall be securely fenced unless it is in such a position or of such construction as to be as safe to every person employed or working on the premises as it would be if securely fenced.

Transmission machinery

13.—(1) Every part of the transmission machinery shall be securely fenced unless it is in such a position or of such construction as to be as safe to every person employed or working on the premises as it would be if securely fenced.

(2) Efficient devices or appliances shall be provided and maintained in every room or place where work is carried on by which the power can promptly be cut off from the transmission machinery in that room or place.

(3) No driving belt when not in use shall be allowed to rest or ride upon a revolving shaft which forms part of the transmission machinery.

(4) Suitable striking gear or other efficient mechanical appliances shall be provided and maintained and used to move driving belts to and from fast and loose pulleys which form part of the transmission machinery, and any such gear or appliances shall be so constructed, placed and maintained as to prevent the driving belt from creeping back on to the fast pulley.

(5) Where the Minister is satisfied that owing to special circumstances the fulfilment of any of the requirements of subsections (2)

to (4) of this section is unnecessary or impracticable, he may by order direct that that requirement shall not apply in those circumstances.

Other machinery

14.—(1) Every dangerous part of any machinery, other than prime movers and transmission machinery, shall be securely fenced unless it is in such a position or of such construction as to be as safe to every person employed or working on the premises as it would be if securely fenced.

(2) In so far as the safety of a dangerous part of any machinery cannot by reason of the nature of the operation be secured by means of a fixed guard, the requirements of subsection (1) of this section shall be deemed to have been complied with if a device is provided which automatically prevents the operator from coming into contact with that part.

(3) (4) [*Repealed from January* 1, 1975, *by the Factories Act* 1961 *etc.* (*Repeals and Modifications*) *Regulations* 1974, *S.I.* 1974 *No.* 1941.]

(5) Any part of a stock-bar which projects beyond the headstock of a lathe shall be securely fenced unless it is in such a position as to be as safe to every person employed or working on the premises as it would be if securely fenced.

(6) [*Repealed by S.I.* 1974 *No.* 1941.]

Provisions as to unfenced machinery

15.—(1) In determining, for the purposes of the foregoing provisions of this Part of this Act, whether any part of machinery is in such a position or of such construction as to be as safe to every person employed or working on the premises as it would be if securely fenced, the following paragraphs shall apply in a case where this section applies, that is to say—

(*a*) no account shall be taken of any person carrying out, while the part of machinery is in motion, an examination thereof or any lubrication or adjustment shown by the examination to be immediately necessary, if the examination, lubrication or adjustment can only be carried out while the part of machinery is in motion; and

(*b*) in the case of any part of transmission machinery used in any such process as may be specified in regulations made by the Minister, being a process where owing to the continuous nature thereof the stopping of that part would seriously interfere with the carrying on of the process, no account shall be taken of any person carrying out, by such methods and in such circumstances as may be specified in the regulations, any lubrication or any mounting or shipping of belts.

(2) This section only applies where the examination, lubrication or other operation is carried out by such [. . .] persons who have attained the age of eighteen as may be specified in regulations made by the Minister, and all such other conditions as may be so specified are complied with.

Construction and maintenance of fencing

16. All fencing or other safeguards provided in pursuance of the foregoing provisions of this Part of this Act shall be of substantial construction, and constantly maintained and kept in position while the parts required to be fenced or safeguarded are in motion or use, except when any such parts are necessarily exposed for examination and for any lubrication or adjustment shown by the examination to be immediately necessary, and all such conditions as may be specified in regulations made by the Minister are complied with.

Floors, passages and stairs

28.—(1) All floors, steps, stairs, passages and gangways shall be of sound construction and properly maintained and shall, so far as is reasonably practicable, be kept free from any obstruction and from any substance likely to cause persons to slip.

(2) For every staircase in a building or affording a means of exit from a building, a substantial hand-rail shall be provided and maintained, which, if the staircase has an open side, shall be on that side, and in the case of a staircase having two open sides or of a staircase which, owing to the nature of its construction or the condition of the surface of the steps or other special circumstances, is specially liable to cause accidents, such a hand-rail shall be provided and maintained on both sides.

(3) Any open side of a staircase shall also be guarded by the provision and maintenance of a lower rail or other effective means.

(4) All openings in floors shall be securely fenced, except in so far as the nature of the work renders such fencing impracticable.

(5) All ladders shall be soundly constructed and properly maintained.

Safe means of access and safe place of employment

29.—(1) There shall, so far as is reasonably practicable, be provided and maintained safe means of access to every place at which any person has at any time to work, and every such place shall, so far as is reasonably practicable, be made and kept safe for any person working there.

(2) Where any person has to work at a place for which he will be liable to fall a distance more than six feet six inches,[4] then, unless the place is one which affords secure foothold and, where necessary, secure hand-hold, means shall be provided, so far as is reasonably practicable, by fencing or otherwise, for ensuring his safety.

Supply of drinking water

57.—(1) There shall be provided and maintained at suitable points conveniently accessible to all persons employed an adequate supply of wholesome drinking water from a public main or from some other source approved in writing by the district council.

(2) A supply of drinking water which is not laid on shall be contained in suitable vessels, and shall be renewed at least daily, and all practicable steps shall be taken to preserve the water and vessels from contamination; and a drinking water supply (whether laid on or not) shall, in such cases as the inspector for the district may direct, be clearly marked "Drinking Water".

(3) Except where the water is delivered in an upward jet from which employed persons can conveniently drink, one or more suitable cups or drinking vessels shall be provided at each point of supply with facilities for rinsing them in drinking water.

(4) The approval required under subsection (1) of this section shall not be withheld except on the ground that the water is not wholesome.

Washing facilities

58.—(1) There shall be provided and maintained for the use of employed persons adequate and suitable facilities for washing which shall include a supply of clean running hot and cold or warm water and, in addition, soap and clean towels or other suitable means of cleaning or drying; and the facilities shall be conveniently accessible and shall be kept in a clean and orderly condition.

(2)–(4) [*Repealed from January* 1, 1975, *by the Factories Act* 1961 *etc.* (*Repeals and Modifications*) *Regulations* 1974 (*S.I.* 1974 *No.* 1941).]

---

[4] 2 metres from 12/8/83.

Accommodation for clothing

59.—(1) There shall be provided and maintained for the use of employed persons adequate and suitable accommodation for clothing not worn during working hours; and such arrangements as are reasonably practicable or, when a standard is prescribed, such arrangements as are laid down thereby shall be made for drying such clothing.

(2) (3). [*Repealed from January* 1, 1975, *by the Factories Act* 1961 *etc.* (*Repeals and Modifications*) *Regulations* 1974 (*S.I.* 1974 *No.* 1941).]

Sitting facilities

60.—(1) Where any employed persons have in the course of their employment reasonable opportunities for sitting without detriment to their work, there shall be provided and maintained for their use suitable facilities for sitting sufficient to enable them to take advantage of those opportunities.

(2) Where a substantial proportion of any work can properly be done sitting—

(*a*) there shall be provided and maintained for any employed person doing that work a seat of a design, construction and dimensions suitable for him and the work, together with a foot-rest on which he can readily and comfortably support his feet if he cannot do so without a foot-rest, and

(*b*) the arrangements shall be such that the seat is adequately and properly supported while in use for the purpose for which it is provided.

(3) For the purposes of subsection (2) of this section the dimensions of a seat which is adjustable shall be taken to be its dimensions as for the time being adjusted.

Part XIV

Interpretation and General

Interpretation

Interpretation of expression "factory"

175.—(1) Subject to the provisions of this section, the expression "factory" means any premises in which, or within the close or curtilage or precincts of which, persons are employed in manual labour in any process for or incidental to any of the following purposes, namely:—

(a) the making of any article or of part of any article; or
(b) the altering, repairing, ornamenting, finishing, cleaning, or washing or the breaking up or demolition of any article; or
(c) the adapting for sale of any article;
(d) the slaughtering of cattle, sheep, swine, goats, horses, asses or mules; or
(e) the confinement of such animals as aforesaid while awaiting slaughter at other premises, in a case where the place of confinement is available in connection with those other premises, is not maintained primarily for agricultural purposes within the meaning of the Agriculture Act, 1947, or, as the case may be, the Agriculture (Scotland) Act, 1948, and does not form part of premises used for the holding of a market in respect of such animals;

being premises in which, or within the close or curtilage or precincts of which, the work is carried on by way of trade or for purposes of gain and to or over which the employer of the persons employed therein has the right of access or control.

(2) The expression "factory" also includes the following premises in which persons are employed in manual labour (whether or not they are factories by virtue of subsection (1) of this section), that is to say,—

(a) any yard or dry dock (including the precincts thereof) in which ships or vessels are constructed, reconstructed, repaired, refitted, finished or broken up;
(b) any premises in which the business of sorting any articles is carried on as a preliminary to the work carried on in any factory or incidentally to the purposes of any factory;
(c) any premises in which the business of washing or filling bottles or containers or packing articles is carried on incidentally to the purposes of any factory;
(d) any premises in which the business of hooking, plaiting, lapping, making-up or packing of yarn or cloth is carried on;
(e) any laundry carried on as ancillary to another business, or incidentally to the purposes of any public institution;
(f) except as provided in subsection (10) of this section, any premises in which the construction, reconstruction or repair of locomotives, vehicles or other plant for use for transport purposes is carried on as ancillary to a transport undertaking or other industrial or commercial undertaking;
(g) any premises in which printing by letterpress, lithography,

photogravure, or other similar process, or bookbinding is carried on by way of trade or for purposes of gain or incidentally to another business so carried on;

(*h*) any premises in which the making, adaptation or repair of dresses, scenery or properties is carried on incidentally to the production, exhibition or presentation by way of trade or for purposes of gain of cinematograph films or theatrical performances, not being a stage or dressing-room of a theatre in which only occasional adaptations or repairs are made;

(*j*) any premises in which the business of making or mending nets is carried on incidentally to the fishing industry;

(*k*) any premises in which mechanical power is used in connection with the making or repair of articles of metal or wood incidentally to any business carried on by way of trade or for purposes of gain;

(*l*) any premises in which the production of cinematograph films is carried on by way of trade or for purposes of gain, so, however, that the employment at any such premises of theatrical performers within the meaning of the Theatrical Employers Registration Act, 1925, and of attendants on such theatrical performers shall not be deemed to be employment in a factory;

(*m*) any premises in which articles are made or prepared incidentally to the carrying on of building operations or works of engineering construction, not being premises in which such operations or works are being carried on;

(*n*) any premises used for the storage of gas in a gasholder having a storage capacity of not less than five thousand cubic feet.[5]

(3) Any line or siding (not being part of a railway or tramway) which is used in connection with and for the purposes of a factory, shall be deemed to be part of the factory; and if any such line or siding is used in connection with more than one factory belonging to different occupiers, the line or siding shall be deemed to be a separate factory.

(4) A part of a factory may, with the approval in writing of the chief inspector, be taken to be a separate factory and two or more factories may, with the like approval, be taken to be a single factory.

(5) Any workplace in which, with the permission of or under agreement with the owner or occupier, two or more persons carry

---

[5] 140 cubic metres from 12/8/83.

on any work which would constitute the workplace a factory if the persons working therein were in the employment of the owner or occupier, shall be deemed to be a factory for the purposes of this Act, and, in the case of any such workplace not being a tenement factory or part of a tenement factory, the provisions of this Act shall apply as if the owner or occupier of the workplace were the occupier of the factory and the persons working therein were persons employed in the factory.

(6) Where a place situate within the close, curtilage, or precincts forming a factory is solely used for some purpose other than the processes carried on in the factory, that place shall not be deemed to form part of the factory for the purposes of this Act, but shall, if otherwise it would be a factory, be deemed to be a separate factory.

(7) Premises shall not be excluded from the definition of a factory by reason only that they are open air premises.

(8) Where the Minister by regulations so directs as respects all or any purposes of this Act, different branches or departments of work carried on in the same factory shall be deemed to be different factories.

(9) Any premises belonging to or in the occupation of the Crown or any municipal or other public authority shall not be deemed not to be a factory, and building operations or works of engineering construction undertaken by or on behalf of the Crown or any such authority shall not be excluded from the operation of this Act, by reason only that the work carried on thereat is not carried on by way of trade or for purposes of gain.

(10) Premises used for the purpose of housing locomotives or vehicles where only cleaning, washing, running repairs or minor adjustments are carried out shall not be deemed to be a factory by reason only of paragraph (f) of subsection (2) of this section, unless they are premises used for the purposes of a railway undertaking where running repairs to locomotives are carried out.

Offices Shops and Railway Premises Act 1963
(c.41)

Sect.

1. Premises to which this Act applies
2. Exception for premises in which only employer's relatives or outworkers work
3. Exception for premises where only 21 man-hours weekly normally worked
4. Cleanliness

5. Overcrowding
6. Temperature
7. Ventilation
8. Lighting
9. Sanitary Conveniences
10. Washing facilities
11. Supply of drinking water
12. Accommodation for clothing
13. Sitting facilities
14. Seats for sedentary work
15. Eating facilities
16. Floors, passages and stairs
17. Fencing of exposed parts of machinery
18. Avoidance of exposure of young persons to danger in cleaning machinery

Scope of Act

Premises to which this Act applies

1.—(1) The premises to which this Act applies are office premises, shop premises and railway premises, being (in each case) premises in the case of which persons are employed to work therein.

(2) In this Act—

(a) "office premises" means a building or part of a building, being a building or part the sole or principal use of which is an office or for office purposes;

(b) "office purposes" includes the purposes of administration, clerical work, handling money and telephone and telegraph operating; and

(c) "clerical work" includes writing, book-keeping, sorting papers, filing, typing, duplicating, machine calculating, drawing and the editorial preparation of matter for publication;

and for the purposes of this Act premises occupied together with office premises for the purposes of the activities there carried on shall be treated as forming part of the office premises.

(3) In this Act—

(a) "shop premises" means—
   (i) a shop;
   (ii) a building or part of a building, being a building or part which is not a shop but of which the sole or principal use is the carrying on there of retail trade or business;

(iii) a building occupied by a wholesale dealer or merchant where goods are kept for sale wholesale or a part of a building so occupied where goods are so kept, but not including a warehouse belonging to the owners, trustees or conservators of a dock, wharf or quay;
(iv) a building to which members of the public are invited to resort for the purpose of delivering their goods for repair or other treatment or of themselves there carrying out repairs to, or other treatment of, goods, or a part of a building to which members of the public are invited to resort for that purpose;
(v) any premises (in this Act referred to as "fuel storage premises") occupied for the purpose of a trade or business which consists of, or includes, the sale of solid fuel, being premises used for the storage of such fuel intended to be sold in the course of that trade or business, but not including dock storage premises or colliery storage premises;

(b) "retail trade or business" includes the sale to members of the public of food or drink for immediate consumption, retail sales by auction and the business of lending books or periodicals for the purpose of gain;
(c) "solid fuel" means coal, coke and any solid fuel derived from coal or of which coal or coke is a constituent;
(d) "dock storage premises" means fuel storage premises which constitute or are comprised in premises to which certain provisions of the Factories Act, 1961, apply by virtue of section 125 (1) (docks, etc.) of that Act; and
(e) "colliery storage premises" means fuel storage premises which form part of premises which, for the purposes of the Mines and Quarries Act, 1954, form part of a mine or quarry, other than premises where persons are regularly employed to work by a person other than the owner (as defined by that Act) of the mine or quarry;

and for the purposes of this Act premises occupied together with a shop or with a building or part of a building falling within sub-paragraph (ii), (iii) or (iv) of paragraph (a) above for the purposes of the trade or business carried on in the shop or, as the case may be, the building or part of a building, shall be treated as forming part of the shop or, as the case may be, of the building or part of the building, and premises occupied together with fuel storage premises for the purposes of the activities there carried on (not being office premises) shall be treated as forming part of the fuel storage premises, but for the purposes of this Act office premises comprised in fuel storage premises shall be deemed not to form part of the last-mentioned premises.

(4) In this Act "railway premises" means a building occupied by railway undertakers for the purposes of the railway undertaking carried on by them and situate in the immediate vicinity of the permanent way or a part (so occupied) of a building so situate, but does not include—

- (*a*) office or shop premises;
- (*b*) premises used for the provision of living accommodation for persons employed in the undertaking, or hotels; or
- (*c*) premises wherein are carried on such processes or operations as are mentioned in section 123 (1) (electrical stations) of the Factories Act, 1961, and for such supply as is therein mentioned.

(5) For the purposes of this Act premises maintained in conjunction with office, shop or railway premises for the purpose of the sale or supply for immediate consumption of food or drink wholly or mainly to persons employed to work in the premises in conjunction with which they are maintained shall, if they neither form part of those premises nor are required by the foregoing provisions of this section to be treated as forming part of them, be treated for the purposes of this Act as premises of the class within which fall the premises in conjunction with which they are maintained.

Exception for premises in which only employer's relatives or outworkers work

2.—(1) This Act shall not apply to any premises to which it would, apart from this subsection, apply, if none of the persons employed to work in the premises is other than the husband, wife, parent, grandparent, son, daughter, grandchild, brother or sister of the person by whom they are so employed.

(2) A dwelling shall not, for the purposes of this Act, be taken to constitute or comprise premises to which this Act applies by reason only that a person dwelling there who is employed by a person who does not so dwell does there the work that he is employed to do in compliance with a term of his contract of service that he shall do it there.

Exception for premises where only 21 man-hours weekly normally worked

3.—(1) This Act shall not apply to any premises to which it would, apart from this subsection, apply, if the period of time worked there during each week does not normally exceed twenty-one hours.

(2) For the purposes of this section the period of time worked in any premises shall be deemed to be—

(a) as regards a week in which one person only is employed to work in the premises, the period of time worked by him there;
(b) as regards a week in which two persons or more are so employed, the sum of the periods of time for which respectively those persons work there.

(3) [*Repealed from January 1, 1975, by the Offices, Shops and Railway Premises Act 1963 (Repeals and Modifications) Regulations 1974 (S.I. 1974 No. 1943).*]

Health, Safety and Welfare of Employees
(General Provisions)

Cleanliness

4.—(1) All premises to which this Act applies, and all furniture, furnishings and fittings in such premises shall be kept in a clean state.

(2) No dirt or refuse shall be allowed to accumulate in any part of premises to which this Act applies in which work, or through which pass, any of the persons employed to work in the premises; and the floors of, and any steps comprised in, any such part as aforesaid shall be cleaned not less than once a week by washing or, if it is effective and suitable, by sweeping or other method.

(3) [*Repealed from January 1, 1975, by the Offices, Shops and Railway Premises Act 1963 (Repeals and Modifications) Regulations 1974 (S.I. 1974 No. 1943).*]

(4) [Subsection (2) of this section shall not] be construed as being in derogation of the general obligation imposed by subsection (1) of this section.

(5) Nothing in this section or in regulations thereunder shall apply to fuel storage premises which are wholly in the open, and, in the case of such premises which are partly in the open, so much of them as is in the open shall, for the purposes of this section and of such regulations, be treated as not forming part of the premises.

Overcrowding

5.—(1) No room comprised in, or constituting, premises to which this Act applies shall, while work is going on therein, be so overcrowded as to cause risk of injury to the health of persons working therein; and in determining, for the purposes of this subsection, whether any such room is so overcrowded as aforesaid, regard shall be had (amongst other things) not only to the number of persons who may be expected to be working in the room at any time but also to the space in the room occupied by furniture, furnishings, fittings, machinery, plant, equipment, appliances and other things (whether similar to any of those aforesaid or not).

(2) The number of persons habitually employed at a time to work in such a room as aforesaid shall not be such that the quotient derived by dividing by that number the number which expresses in [square metres] the area of the surface of the floor of the room is less than [3·7] or the quotient derived by dividing by the first-mentioned number the number which expresses in [cubic metres] the capacity of the room is less than [11].

(3) Subsection (2) of this section—

(*a*) shall not prejudice the general obligation imposed by subsection (1) thereof;
(*b*) shall not apply to a room to which members of the public are invited to resort; and
(*c*) shall not, in the case of a room comprised in, or constituting, premises of any class (being a room which at the passing of this Act is comprised in, or constitutes, premises to which this Act applies), have effect until the expiration of the period of three years beginning with the day on which the said subsection (1) comes into force as respects premises of that class.

Temperature

6.—(1) Effective provision shall be made for securing and maintaining a reasonable temperature in every room comprised in, or constituting, premises to which this Act applies, being a room in which persons are employed to work otherwise than for short periods, but no method shall be used which results in the escape into the air of any such room of any fume of such a character and to such extent as to be likely to be injurious or offensive to persons working therein.

(2) Where a substantial proportion of the work done in a room to which the foregoing subsection applies does not involve severe physical effort, a temperature of less than [16 degrees Celsius] shall not be deemed, after the first hour, to be a reasonable temperature while work is going on.

(3) The foregoing subsections shall not apply—

(a) to a room which comprises, or is comprised in or constitutes, office premises, being a room to which members of the public are invited to resort, and in which the maintenance of a reasonable temperature is not reasonably practicable; or
(b) to a room which comprises, or is comprised in or constitutes, shop or railway premises, being a room in which the maintenance of a reasonable temperature is not reasonably practicable or would cause deterioration of goods;

but there shall be provided for persons who are employed to work in a room to which, but for the foregoing provisions of this subsection, subsection (1) of this section would apply, conveniently accessible and effective means of enabling them to warm themselves.

(4) In premises to which this Act applies there shall, on each floor on which there is a room to which subsection (1) of this section applies, be provided in a conspicuous place and in such a position as to be easily seen by the persons employed to work in the premises on that floor a thermometer of a kind suitable for enabling the temperature in any such room on that floor to be readily determined; and a thermometer provided in pursuance of this subsection shall be kept available for use by those persons for that purpose.

(5) [*Repealed from January* 1, 1975, *by the Offices, Shops and Railway Premises Act* 1963 (*Repeals and Modifications*) *Regulations* 1974 (*S.I.* 1974 *No.* 1943).]

(6) It shall be the duty of the employer of persons for whom means of enabling them to warm themselves are provided in pursuance of subsection (3) of this section to afford them reasonable opportunities for using those means, and if he fails so to do he shall be guilty of an offence.

(7) In this section "fume" includes gas or vapour.

Ventilation

7.—(1) Effective and suitable provision shall be made for securing and maintaining, by the circulation of adequate supplies of fresh or artificially purified air, the ventilation of every room comprised in, or constituting, premises to which this Act applies, being a room in which persons are employed to work.

(2) [*Repealed from January* 1, 1975, *by the Offices, Shops and Railway Premises Act* 1963 (*Repeals and Modifications*) *Regulations* 1974 (*S.I.* 1974 *No.* 1943).]

Lighting

8.—(1) Effective provision shall be made for securing and maintaining, in every part of premises to which this Act applies in which persons are working or passing, sufficient and suitable lighting, whether natural or artificial.

(2) [*Repealed from January 1, 1975, by the Offices, Shops and Railway Premises Act 1963 (Repeals and Modifications) Regulations 1974 (S.I. 1974 No. 1943).*]

(3) All glazed windows and skylights used for the lighting of any part of premises to which this Act applies in which work, or through which pass, any of the persons employed to work in the premises shall, so far as reasonably practicable, be kept clean on both the inner and outer surfaces and free from obstruction; but this subsection shall not affect the white-washing or shading of windows or skylights for the purpose of mitigating heat or glare.

(4) All apparatus installed at premises to which this Act applies for producing artificial lighting thereat in parts in which the securing of lighting is required by this section to be provided for shall be properly maintained.

Sanitary conveniences

9.—(1) There shall, in the case of premises to which this Act applies, be provided, at places conveniently accessible to the persons employed to work in the premises, suitable and sufficient sanitary conveniences for their use.

(2) Conveniences provided in pursuance of the foregoing subsection shall be kept clean and properly maintained and effective provision shall be made for lighting and ventilating them.

(3), (4) [*Repealed from January 1, 1975, by the Offices, Shops and Railway Premises Act 1963 (Repeals and Modifications) Regulations 1974 (S.I. 1974 No. 1943).*]

(5) Subsection (1) of this section shall be deemed to be complied with in relation to any premises as regards any period during which there are in operation arrangements for enabling the persons employed to work in the premises to have the use of sanitary conveniences provided for the use of others, being conveniences whose provision would have constituted compliance with that subsection had they been provided in pursuance thereof for the first-mentioned persons and with respect to which the requirements of subsection (2) of this section are satisfied.

(6) Neither [section 45] of the Public Health Act, 1936, nor section 29 of the Public Health (Scotland) Act, 1897, nor section 106 of the Public Health (London) Act, 1936 (which relate to the pro-

visions and repair of sanitary conveniences for factories, &c.), shall apply to premises to which this Act applies.

Washing facilities

10.—(1) There shall, in the case of premises to which this Act applies, be provided, at places conveniently accessible to the persons employed to work in the premises, suitable and sufficient washing facilities, including a supply of clean, running hot and cold or warm water and, in addition, soap and clean towels or other suitable means of cleaning or drying.

(2) Every place where facilities are provided in pursuance of this section shall be provided with effective means of lighting it and be kept clean and in orderly condition, and all apparatus therein for the purpose of washing or drying shall be kept clean and be properly maintained.

(3), (4) [*Repealed from January* 1, 1975, *by the Offices, Shops and Railway Premises Act* 1963 (*Repeals and Modifications*) *Regulations* 1974 (*S.I.* 1974 *No.* 1943).]

(5) Subsection (1) of this section shall be deemed to be complied with in relation to any premises as regards any period during which there are in operation arrangements for enabling the persons employed to work in the premises to have the use of washing facilities provided for the use of others, being facilities whose provision would have constituted compliance with that subsection had they been provided in pursuance thereof for the first-mentioned persons and which are provided at a place with respect to which the requirements of subsection (2) of this section are satisfied.

Supply of drinking water

11.—(1) There shall, in the case of premises to which this Act applies, be provided and maintained, at suitable places conveniently accessible to the persons employed to work in the premises, an adequate supply of wholesome drinking water.

(2) Where a supply of water provided at a place in pursuance of the foregoing subsection is not piped, it must be contained in suitable vessels and must be renewed at least daily; and all practicable steps must be taken to preserve it and the vessels in which it is contained from contamination.

(3) Where water a supply of which is provided in pursuance of this section is delivered otherwise than in a jet from which persons can conveniently drink, there shall either—

(a) be provided, and be renewed so often as occasion requires, a supply of drinking vessels of a kind designed to be discarded after use; or

(b) be provided a sufficient number of drinking vessels of a kind other than as aforesaid, together with facilities for rinsing them in clean water.

(4) Subsection (1) of this section shall be deemed to be complied with in relation to any premises as regards any period during which there are in operation arrangements for enabling the persons employed to work in the premises to avail themselves of a supply of drinking water provided and maintained for the use of others, being a supply whose provision and maintenance would have constituted compliance with that subsection had it been provided and maintained for the use of the first-mentioned persons, and—

(a) where the supply provided is not piped, the requirements of subsection (2) of this section are satisfied as respects it and the vessels in which it is contained; and

(b) where the water supplied is delivered as mentioned in subsection (3) of this section, the requirements of that subsection are satisfied.

Accommodation for clothing

12.—(1) There shall, in the case of premises to which this Act applies,—

(a) be made, at suitable places, suitable and sufficient provision for enabling such of the clothing of the persons employed to work in the premises as is not worn by them during working hours to be hung up or otherwise accommodated; and

(b) be made, for drying that clothing, such arrangements as are reasonably practicable or, if a standard of arrangements for drying that clothing is prescribed, such arrangements as conform to that standard.

(2) Where persons are employed to do such work in premises to which this Act applies as necessitates the wearing of special clothing, and they do not take that clothing home, there shall, in the case of those premises,—

(a) be made, at suitable places, suitable and sufficient provision for enabling that clothing to be hung up or otherwise accommodated; and

(b) be made, for drying that clothing, such arrangements as are reasonably practicable or, if a standard of arrangements for drying that clothing is prescribed, such arrangements as conform to that standard.

(3) [*Repealed from January* 1, 1975, *by the Offices, Shops and Railway Premises Act* 1963 *(Repeals and Modifications) Regulations* 1974 *(S.I.* 1974 *No.* 1943).]

Sitting facilities

13.—(1) Where persons who are employed to work in office, shop or railway premises have, in the course of their work, reasonable opportunities for sitting without detriment to it, there shall be provided for their use, at suitable places conveniently accessible to them, suitable facilities for sitting sufficient to enable them to take advantage of those opportunities.

(2) Where persons are employed to work in a room which comprises, or is comprised in or constitutes, shop premises, being a room whereto customers are invited to resort, and have in the course of their work, reasonable opportunities for sitting without detriment to it, facilities provided for their use in pursuance of subsection (1) of this section shall be deemed not to be sufficient if the number of seats provided and the number of the persons employed are in less ratio than 1 to 3.

(3) It shall be the duty of the employer of persons for whose use facilities are provided in pursuance of the foregoing provisions of this section to permit them to use them whenever the use thereof does not interfere with their work, and if he fails so to do he shall be guilty of an offence.

Seats for sedentary work

14.—(1) Without prejudice to the general obligation imposed by the last foregoing section, where any work done in any premises to which this Act applies is of such a kind that it (or a substantial part of it) can, or must, be done sitting, there shall be provided for each person employed to do it there a seat of a design, construction and dimensions suitable for him and it, together with a foot-rest on which he can readily and comfortably support his feet if he cannot do so without one.

(2) A seat provided in pursuance of the foregoing subsection, and a foot-rest so provided that does not form part of a seat, must be adequately and properly supported while in use for the purposes for which it is provided.

(3) For the purpose of subsection (1) of this section, the dimensions of an adjustable seat shall be taken to be its dimensions as for the time being adjusted.

Eating facilities

15. Where persons employed to work in shop premises eat meals there, suitable and sufficient facilities for eating them shall be provided.

Floors, passages and stairs

16.—(1) All floors, stairs, steps, passages and gangways comprised in premises to which this Act applies shall be of sound construction and properly maintained and shall, so far as is reasonably practicable, be kept free from obstruction and from any substance likely to cause persons to slip.
(2) For every staircase comprised in such premises as aforesaid, a substantial hand-rail or hand-hold shall be provided and maintained, which, if the staircase has an open side, shall be on that side; and in the case of a staircase having two open sides or of a staircase which, owing to the nature of its construction or the condition of the surface of the steps or other special circumstances, is specially liable to cause accidents, such a hand-rail or hand-hold shall be provided and maintained on both sides.
(3) Any open side of a staircase to which the last foregoing subsection applies, shall also be guarded by the provision and maintenance of efficient means of preventing any person from accidentally falling through the space between the hand-rail or hand-hold and the steps of the staircase.
(4) All openings in floors comprised in premises to which this Act applies shall be securely fenced, except in so far as the nature of the work renders such fencing impracticable.
(5) The foregoing provisions of this section shall not apply to any such part of any fuel storage premises as is in the open, but in relation to any such part the following provisions shall have effect, namely,—

   (*a*) the surface of the ground shall be kept in good repair;
   (*b*) all steps and platforms shall be of sound construction and properly maintained;
   (*c*) all openings in platforms shall be securely fenced, except in so far as the nature of the work renders such fencing impracticable.

Fencing of exposed parts of machinery

17.—(1) Every dangerous part of any machinery used as, or forming, part of the equipment of premises to which this Act applies shall be securely fenced unless it is in such a position or of such construction as to be as safe to every person working in the premises as it would be if securely fenced.

(2) In so far as the safety of a dangerous part of any machinery cannot, by reason of the nature of the operation effected by means of the machinery, be secured by means of a fixed guard, the requirements of the foregoing subsection shall be deemed to be complied with if a device is provided that automatically prevents the operator from coming into contact with that part.

(3) In determining, for the purposes of subsection (1) of this section, whether a moving part of any machinery is in such a position or of such construction as is therein mentioned, no account shall be taken of any person carrying out while the part is in motion an examination thereof or any lubrication or adjustment shown by the examination to be immediately necessary, if the examination, lubrication or adjustment can only be carried out while the part is in motion.

(4) Fencing provided in pursuance of the foregoing provisions of this section shall be of substantial construction, be properly maintained and be kept in position while the parts required to be fenced are in motion or use, except when any such parts are necessarily exposed for examination and for any lubrication or adjustment shown by the examination to be immediately necessary.

(5) Subsection (3) of this section, and so much of subsection (4) thereof as relates to the exception from the requirement thereby imposed, shall only apply where the examination, lubrication or adjustment in question is carried out by such persons who have attained the age of eighteen as may be specified in regulations made by the Minister and all other such conditions as may be so specified are complied with.

Avoidance of exposure of young persons to danger in cleaning machinery

18.—(1) No young person employed to work in premises to which this Act applies shall clean any machinery used as, or forming, part of the equipment of the premises if doing so exposes him to risk of injury from a moving part of that or any adjacent machinery.

(2) In this section "young person" means a person who has not attained the age of eighteen.

Health and Safety at Work etc. Act 1974
(c. 37)

Sect.

1. Preliminary
2. General duties of employers to their employees
3. General duties of employers and self-employed to persons other than their employees
4. General duties of persons concerned with premises to persons other than their employees
5. General duty of persons in control of certain premises in relation to harmful emissions into atmosphere
6. General duties of manufacturers etc. as regards articles and substances for use at work
7. General duties of employees at work
8. Duty not to interfere with or misuse things provided pursuant to certain provisions
9. Duty not to charge employees for things done or provided pursuant to certain specific enactments
47. Civil liability

Preliminary

Preliminary

1.—(1) The provisions of this Part shall have effect with a view to—

- (*a*) securing the health, safety and welfare of persons at work;
- (*b*) protecting persons other than persons at work against risks to health or safety arising out of or in connection with the activities of persons at work;
- (*c*) controlling the keeping and use of explosive or highly flammable or otherwise dangerous substances, and generally preventing the unlawful acquisition, possession and use of such substances; and
- (*d*) controlling the emission into the atmosphere of noxious or offensive substances from premises of any class prescribed for the purposes of this paragraph.

(2) The provisions of this Part relating to the making of health and safety regulations [. . .] and the preparation and approval of codes of practice shall in particular have effect with a view to enabling the enactments specified in the third column of Schedule 1 and the regulations, orders and other instruments in force under those enactments to be progressively replaced by a system of regulations and approved codes of practice operating in combination

with the other provisions of this Part and designed to maintain or improve the standards of health, safety and welfare established by or under those enactments.

(3) For the purposes of this Part risks arising out of or in connection with the activities of persons at work shall be treated as including risks attributable to the manner of conducting an undertaking, the plant or substances used for the purposes of an undertaking and the condition of premises so used or any part of them.

(4) References in this Part to the general purposes of this Part are references to the purposes mentioned in subsection (1) above.

General duties

General duties of employers to their employees

2.—(1) It shall be the duty of every employer to ensure, so far as is reasonably practicable, the health, safety and welfare at work of all his employees.

(2) Without prejudice to the generality of an employer's duty under the preceding subsection, the matters to which that duty extends include in particular—

- (*a*) the provision and maintenance of plant and systems of work that are, so far as is reasonably practicable, safe and without risks to health;
- (*b*) arrangements for ensuring, so far as is reasonably practicable, safety and absence of risks to health in connection with the use, handling, storage and transport of articles and substances;
- (*c*) the provision of such information, instruction, training and supervision as is necessary to ensure, so far as is reasonably practicable, the health and safety at work of his employees;
- (*d*) so far as is reasonably practicable, as regards any place of work under the employer's control, the maintenance of it in a condition that is safe and without risks to health and the provision and maintenance of means of access to and egress from it that are safe and without such risks;
- (*e*) the provision and maintenance of a working environment for his employees that is, so far as is reasonably practicable, safe, without risks to health, and adequate as regards facilities and arrangements for their welfare at work.

(3) Except in such cases as may be prescribed, it shall be the duty of every employer to prepare and as often as may be appropriate revise a written statement of his general policy with respect to the health and safety at work of his employees and the organis-

ation and arrangements for the time being in force for carrying out that policy and to bring the statement and any revision of it to the notice of all of his employees.

(4) Regulations made by the Secretary of State may provide for the appointment in prescribed cases by recognised trade unions (within the meaning of the regulations) of safety representatives from amongst the employees, and those representatives shall represent the employees in consultations with the employers under subsection (6) below and shall have such other functions as may be prescribed.

(5) [. . .]

(6) It shall be the duty of every employer to consult any such representatives with a view to the making and maintenance of arrangements which will enable him and his employees to co-operate effectively in promoting and developing measures to ensure the health and safety at work of the employees, and in checking the effectiveness of such measures.

(7) In such cases as may be prescribed it shall be the duty of every employer, if requested to do so by the safety representatives mentioned in [subsection (4)] [. . .] above, to establish, in accordance with regulations made by the Secretary of State, a safety committee having the function of keeping under review the measures taken to ensure the health and safety at work of his employees and such other functions as may be prescribed.

General duties of employers and self-employed to persons other than their employees

3.—(1) It shall be the duty of every employer to conduct his undertaking in such a way as to ensure, so far as is reasonably practicable, that persons not in his employment who may be affected thereby are not thereby exposed to risks to their health or safety.

(2) It shall be the duty of every self-employed person to conduct his undertaking in such a way as to ensure, so far as is reasonably practicable, that he and other persons (not being his employees) who may be affected thereby are not thereby exposed to risks to their health or safety.

(3) In such cases as may be prescribed, it shall be the duty or every employer and every self-employed person, in the prescribed circumstances and in the prescribed manner, to give to persons (not being his employees) who may be affected by the way in which he conducts his undertaking the prescribed information about such aspects of the way in which he conducts his undertaking as might affect their health or safety.

General duties of persons concerned with premises to persons other than their employees

4.—(1) This section has effect for imposing on persons duties in relation to those who—

(*a*) are not their employees; but
(*b*) use non-domestic premises made available to them as a place of work or as a place where they may use plant or substances provided for their use there,

and applies to premises so made available and other non-domestic premises used in connection with them.

(2) It shall be the duty of each person who has, to any extent, control of premises to which this section applies or of the means of access thereto or egress therefrom or of any plant or substance in such premises to take such measures as it is reasonable for a person in his position to take to ensure, so far as is reasonably practicable, that the premises, all means of access thereto or egress therefrom available for use by persons using the premises, and any plant or substance in the premises or, as the case may be, provided for use there, is or are safe and without risks to health.

(3) Where a person has, by virtue of any contract or tenancy, an obligation of any extent in relation to—

(*a*) the maintenance or repair of any premises to which this section applies or any means of access thereto or egress therefrom; or
(*b*) the safety of or the absence of risks to health arising from plant or substances in any such premises;

that person shall be treated, for the purposes of subsection (2) above, as being a person who has control of the matters to which his obligation extends.

(4) Any reference in this section to a person having control of any premises or matter is a reference to a person having control of the premises or matter in connection with the carrying on by him of a trade, business or other undertaking (whether for profit or not).

General duty of persons in control of certain premises in relation to harmful emissions into atmosphere

5.—(1) It shall be the duty of the persons having control of any premises of a class prescribed for the purposes of section 1(1)(*d*) to use the best practicable means for preventing the emission into the atmosphere, from the premises, of noxious or offensive substances and for rendering harmless and inoffensive such substances as may be so emitted.

(2) The reference in subsection (1) above to the means to be used for the purposes there mentioned includes a reference to the manner in which the plant provided for those purposes is used and to the supervision of any operation involving the emission of the substances to which that subsection applies.

(3) Any substance or a substance of any description prescribed for the purposes of subsection (1) above as noxious or offensive shall be a noxious or, as the case may be, an offensive substance for those purposes whether or not it would be so apart from this subsection.

(4) Any reference in this section to a person having control of any premises is a reference to a person having control of the premises in connection with the carrying on by him of a trade, business or other undertaking (whether for profit or not) and any duty imposed on any such person by this section shall extend only to matters within his control.

General duties of manufacturers etc. as regards articles and substances for use at work

6.—(1) It shall be the duty of any person who designs, manufactures, imports or supplies any article for use at work or any article of fairground equipment—

- (*a*) to ensure, so far as is reasonably practicable, that the article is so designed and construed that it will be safe and without risks to health at all times when it is being set, used, cleaned or maintained by a person at work;
- (*b*) to carry out or arrange for the carrying out of such testing and examination as may be necesssary for the performance of the duty imposed on him by the preceding paragraph;
- (*c*) to take such steps as are necessary to secure that persons supplied by that person with the article are provided with adequate information about the use for which the article is designed or has been tested and about any conditions necessary to ensure that it will be safe and without risks to health at all such times as are mentioned in paragraph (*a*) above and when it is being dismantled or disposed of; and
- (*d*) to take such steps as are necessary to secure, so far as is reasonably practicable, that persons so supplied are provided with all such revisions of information provided to them by virtue of the preceding paragraph as are necessary

by reason of its becoming known that anything gives rise to a serious risk to health or safety.

(1A) It shall be the duty of any person who designs, manufactures, imports or supplies any article of fairground equipment—
- (*a*) to ensure, so far as is reasonably practicable, that the article is so designed and constructed that it will be safe and without risks to health at all times when it is being used for or in connection with the entertainment of members of the public;
- (*b*) to carry out or arrange for the carrying out of such testing and examination as may be necessary for the performance of the duty imposed on him by the preceding paragraph;
- (*c*) to take such steps as are necessary to secure that persons supplied by that person with the article are provided with adequate information about the use for which the article is designed or has been tested and about any conditions necessary to ensure that it will be safe and without risks to health at all times when it is being used for or in connection with the entertainment of members of the public; and
- (*d*) to take such steps as are necessary to secure, so far as is reasonably practicable, that persons so supplied are provided with all such revisions of information provided to them by virtue of the preceding paragraph as are necessary by reason of its becoming known that anything gives rise to a serious risk to health or safety.

(2) It shall be the duty of any person who undertakes the design or manufacture of any article for use at work [or of any article of fairground equipment] to carry out or arrange for the carrying out of any necessary research with a view to the discovery and, so far as is reasonably practicable, the elimination or minimisation of any risks to health or safety to which the design or article may give rise.

(3) It shall be the duty of any person who erects or installs any article for use at work in any premises where that article is to be used by persons at work [or who erects or installs any article of fairground equipment] to ensure, so far as is reasonably practicable, that nothing about the way in which [the article is erected or installed makes it unsafe or a risk to health at such time as is mentioned in paragraph (a) of subsection (1) or, as the case may be, in paragraph (*a*) of subsection (1) or (1A) above].

(4) It shall be the duty of any person who manufactures, imports or supplies any substance—
- (*a*) to ensure, so far as is reasonably practicable, that the substance will be safe and without risk to health at all times

when it is being used, handled, processed, stored or transported by a person at work or in premises to which section 4 above applies;
(b) to carry out or arrange for the carrying out of such testing and examination as may be necessary for the performance of the duty imposed on him by the preceding paragraph;
(c) to take such steps as are necessary to secure that persons supplied by that person with the substance are provided with adequate information about any risks to health or safety to which the inherent properties of the substance may give rise, about the results of any relevant tests which have been carried out on or in connection with the substance and about any conditions necessary to ensure that the substance will be safe and without risks to health at all such times as are mentioned in paragraph (a) above and when the substance is being disposed of; and
(d) to take such steps as are necessary to secure, so far as is reasonably practicable, that persons so supplied are provided with all such revisions of information provided to them by virtue of the preceding paragraph as are necessary by reason of its becoming known that anything gives rise to a serious risk to health or safety].

(5) It shall be the duty of any person who undertakes the manufacture of any [substance] carry out or arrange for the carrying out of any necessary research with a view to the discovery and, so far as is reasonably practicable, the elimination or minimisation of any risks to health or safety to which the substance may give rise [at all such times as are mentioned in paragraph (a) of subsection (4) above].

(6) Nothing in the preceding provisions of this section shall be taken to require a person to repeat any testing, examination or research which has been carried out otherwise than by him or at his instance, in so far as it is reasonable for him to rely on the results thereof for the purposes of those provisions.

(7) Any duty imposed on any person by any of the preceding provisions of this section shall extend only to things done in the course of a trade, business or other undertaking carried on by him (whether for profit or not) and to matters within his control.

(8) Where a person designs, manufactures, imports or supplies an article [for use at work or an article of fairground equipment and does so for or to another] on the basis of a written undertaking by that other to take specified steps sufficient to ensure, so far as is reasonably practicable, that the article will be safe and without risks to health [at all such times as are mentioned in paragraph (a)

of subsection (1) or, as the case may be, in paragraph (a) of subsection (1) or (1A) above], the undertaking shall have the effect of relieving the first-mentioned person from the duty imposed [by virtue of that paragraph] to such extent as is reasonable having regard to the terms of the undertaking.

[(8A) Nothing in subsection (7) or (8) above shall relieve any person who imports any article or substance from any duty in respect of anything which—

(*a*) in the case of an article designed outside the United Kingdom, was done by and in the course of any trade, profession or other undertaking carried on by, or was within the control of, the person who designed the article; or

(*b*) in the case of an article or substance manufactured outside the United Kingdom, was done by and in the course of any trade, profession or other undertaking carried on by, or was within the control of, the person who manufactured the article or substance.]

(9) Where a person ("the ostensible supplier") supplies any [article or substance] to another ("the customer") under a hire-purchase agreement, conditional sale agreement or credit-sale agreement, and the ostensible supplier—

(*a*) carries on the business of financing the acquisition of goods by others by means of such agreements; and

(*b*) in the course of that business acquired his interest in the article or substance supplied to the customer as a means of financing its acquisition by the customer from a third person ("the effective supplier"),

the effective supplier and not the ostensible supplier shall be treated for the purposes of this section as supplying the article or substance to the customer, and any duty imposed by the preceding provisions of this section on suppliers shall accordingly fall on the effective supplier and not on the ostensible supplier.

[(10) For the purposes of this section absence of safety or a risk to health shall be disregarded in so far as the case in or in relation to which it would arise is shown to be one the occurrence of which could not reasonably be foreseen; and in determining whether any duty imposed by virtue of paragraph (*a*) of subsection (1), (1A) or (4) above has been performed regard shall be had to any relevant information or advice which has been provided to any person by the person by whom the article has been designed, manufactured, imported or supplied or, as the case may be, by the person by whom the substance has been manufactured, imported or supplied.]

General duties of employees at work

7. It shall be the duty of every employee while at work—
   (a) to take reasonable care for the health and safety of himself and of other persons who may be affected by his acts or omissions at work; and
   (b) as regards any duty or requirement imposed on his employer or any other person by or under any of the relevant statutory provisions, to co-operate with him so far as is necessary to enable that duty or requirement to be performed or complied with.

Duty not to interfere with or misuse things provided pursuant to certain provisions

8. No person shall intentionally or recklessly interfere with or misuse anything provided in the interests of health, safety or welfare in pursuance of any of the relevant statutory provisions.

Duty not to charge employees for things done or provided pursuant to certain specific enactments

9. No employer shall levy or permit to be levied on any employee of his any charge in respect of anything done or provided in pursuance of any specific requirement of the relevant statutory provisions.

Civil liability

47.—(1) Nothing in this Part shall be construed—
   (a) as conferring a right of action in any civil proceedings in respect of any failure to comply with any duty imposed by sections 2 to 7 or any contravention of section 8; or
   (b) as affecting the extent (if any) to which breach of a duty imposed by any of the existing statutory provisions is actionable; or
   (c) as affecting the operation of section 12 of the Nuclear Installations Act 1965 (right to compensation by virtue of certain provisions of that Act).

(2) Breach of a duty imposed by health and safety regulations [. . .] shall, so far as it causes damage, be actionable except in so far as the regulations provide otherwise.

(3) No provision made by virtue of section 15 (6) (b) shall afford a defence in any civil proceedings, whether brought by virtue of

subsection (2) above or not; but as regards any duty imposed as mentioned in subsection (2) above health and safety regulations [. . .] may provide for any defence specified in the regulations to be available in any action for breach of that duty.

(4) Subsections (1)(*a*) and (2) above are without prejudice to any right of action which exists apart from the provisions of this Act, and subsection (3) above is without prejudice to any defence which may be available apart from the provisions of the regulations there mentioned.

(5) Any term of an agreement which purports to exclude or restrict the operation of subsection (2) above, or any liability arising by virtue of that subsection, shall be void, except in so far as health and safety regulations [. . .] provide otherwise.

(6) In this section "damage" includes the death of, or injury to, any person (including any disease and any impairment of a person's physical or mental condition).

Appendix 2

PRINCIPAL STATUTORY INSTRUMENTS

General Orders and Regulations

	S.I. No:
Asbestos Regulations	1969/609
Asbestos (Licensing) Regulations	1983/1649
Asbestos (Prohibitions) Regulations	1985/910
Building (Safety, Health and Welfare) Regulations	1948/1145
Carcinogenic Substances Regulations	1967/879
Chemical Works Regulations	1922/731
Classification and Labelling of Explosives Regulations	1983/1140
Classification, Packaging and Labelling of Dangerous Substances Regulations	1984/1244
Construction (General Provisions) Regulations	1961/1580
Construction (Health & Welfare) Regulations	1966/95
Construction (Lifting Operations) Regulations	1961/1581
Construction (Working Places) Regulations	1966/94
Control of Lead at Work Regulations	1980/1248
Control of Major Accident Hazards Regulations	1984/1902
Dangerous Substances (Conveyance by Road in Road Tankers and Tank Containers) Regulations	1981/1059
Dangerous Substances in Harbour Areas Regulations	1987/37
Diving Operations at Work Regulations	1981/399
Docks Regulations	1925/231
Docks Regulations	1934/279

Engineering Construction (Extension of Definition) Regulations	1960/421
Factories (Cleanliness of Walls and Ceilings) Order	1960/1794
Fire Precautions Act 1971 (Modification) Regulations	1976/2007
Fire Precautions (Application for Certificate) Regulations	1976/2008
Fire Precautions (Factories, Offices, Shops and Railway Premises) Order	1976/2009
Fire Precautions (Non-certificated Factory, Office Shop and Railway Premises) Regulations	1976/2010
Grinding of Metals (Miscellaneous Industries) Regulations	1925/904
Health and Safety (Dangerous Pathogens) Regulations	1981/1113
Health and Safety (Emissions into the Atmosphere) Regulations	1983/943
Health and Safety (Enforcing Authority) Regulations	1977/746
Health and Safety (First Aid) Regulations	1981/917
Health and Safety (Genetic Manipulation) Regulations	1978/752
Health and Safety at Work etc. Act 1974 (Application Outside Great Britain) Order	1977/1232
Highly Flammable Liquids and Liquefied Petroleum Gases Regulations	1972/917
Ionising Radiations Regulations	1985/1333
Iron and Steel Foundries Regulations	1953/1464
Notification of Installations Handling Hazardous Substances Regulations	1982/1357
Notification of New Substances Regulations	1982/1496
Operations at Unfenced Machinery Regulations	1938/841
Pneumoconiosis etc. (Workers Compensation) (Determination of Claims) Regulations	1985/1645
Pneumoconiosis etc. (Workers Compensation) (Payment of Claims) Regulations	1985/2035
Pneumoconiosis etc. (Workers Compensation) (Specified Diseases) Order	1985/2034

Pottery (Health) Special Regulations	1947/2161
Protection of Eyes Regulations	1974/1681
Reporting of Industrial Diseases and Dangerous Occurrences Regulations	1985/2023
Road Traffic (Carriage of Dangerous Substances etc.) Regulations	1986/1951
Safety Representatives and Safety Committees Regulations	1977/500
Safety Signs Regulations	1980/1471
Shipbuilding and Ship-repairing Regulations	1960/1932
Social Security (Industrial Injuries) (Prescribed Diseases) Regulations	1985/967
Washing Facilities (Miscellaneous Industries) Regulations	1960/1214
Washing Facilities (Running Water) Exemption Regulations	1960/1029
Woodworking Machines Regulations	1974/903
Work in Compressed Air Special Regulations	1958/61

Offices, Shops and Railway Premises Orders and Regulations

Information for Employees Regulations	1965/307
Offices, Shops and Railway Premises Act 1963 (Exemption No. 1) Order	1964/964
Offices, Shops and Railway Premises Act 1963 (Exemption No. 7) Order	1968/1947
Offices, Shops and Railway Premises Act 1963 (Exemption No. 8) Order	1969/1323
Offices, Shops and Railway Premises Act 1963 (Exemption No. 10) Order	1972/1086
Offices Shops and Railway Premises (Hoists and Lifts) Regulations	1968/849
Prescribed Dangerous Machines Order	1964/971
Sanitary Conveniences Regulations	1964/966
Washing Facilities Regulations	1965/965

Agriculture Regulations

Agriculture (Avoidance of Accidents to Children) Regulations	1958/366

Agriculture (Circular Saw) Regulations	1859/427
Agriculture (Field Machinery) Regulations	1962/1472
Agriculture (Ladders) Regulations	1957/1385
Agriculture (Lifting of Heavy Weights) Regulations	1959/2120
Agriculture (Power Take-off) Regulations	1957/1386
Agriculture (Safeguarding of Workplaces) Regulations	1959/428
Agriculture (Stationery Machinery) Regulations	1959/1216
Agriculture (Threshers and Balers) Regulations	1960/1199
Agriculture (Tractor Cabs) Regulations	1974/2034
Poisonous Substances in Agriculture Regulations	1984/1114

Index

ACCESS,
   employers duty to provide safe, 27
ADJUDICATION,
   social security issues on, 108
AGENTS,
   liability for acts of,
      cases of negligence in, 37
AGRICULTURE,
   definition, 2
APPEALS,
   industrial tribunals to,
      improvement and prohibition notices against, 115
ARMED FORCES,
   members as,
      employees as, 8

BUILDING REGULATIONS,
   safety of place of work, 28

CARE,
   standard of,
      claims of tort in, 12
CAUSATION,
   cases of negligence in, 22
CLAIMS,
   contribution of damages for, 8
   diseases and personal injuries for,
      damages for, 5
   diseases for, 5. *See also* DISEASES.
   employees by,
      courts in, 5 *et seq.*
   indemnity, of, 8
   personal injuries for, 5. *See also* PERSONAL INJURIES.
COMMON EMPLOYMENT DOCTRINE, 33
COMMON LAW,
   claims at,
      relevance of statutory provisions to, 18
   definition of claim at, 5
COMPULSORY INSURANCE,
   cases of negligence in, 55

CONSENT, 36
CONTRACTS,
   employment of, 56, 57
CONTRIBUTORY NEGLIGENCE,
   defences in tort, 34
   degree, 34
   minor of, 35
   prevailing practice, 36
   statutory duty breach of, 77
CONSTRUCTIVE DISMISSAL,
   claims for,
      industrial tribunal at, 98
COURTS,
   claims in, 5 *et seq.*
CRIMINAL ACT,
   vicarious liability for,
      cases of negligence in, 40
CRIMINAL LAW,
   penalties under,
      non-compliance of safety, health and welfare legislation for, 115
CROWN,
   application to,
      occupiers's liability, of, 84
   defendant as, 7

DAMAGES,
   disclosure of documents for,
      cases of negligence in, 43
   exclusion clauses for,
      negligence cases in, 51
   fatal accidents for, 88, 92
   interest on,
      personal injury cases in, 50
   limitation period for,
      cases of negligence in, 52
   loss of amenities for,
      cases of negligence in, 46
   loss of earnings,
      cases of negligence in, 45
   loss of medical expensers,
      cases of negligence in, 45

DAMAGES—*cont.*
  object of,
    cases of negligence in, 43
  pain and suffering for,
    negligences, cases in, 46
  quantification of,
    tort of negligence in, 45
  remoteness of,
    tort of negligence in, 44
  social security and,
    negligence cases in, ???
  statutory duty breach of, for, 78
  taxation and,
    negligence cases in, 48
  time limit,
    dismissal for want of prosecution, 54
DEAFNESS,
  prescribed disease as, 32
DEFENCES,
  cases of negligence in,
    contributory negligence, 34
    volenti non fit injuria, 36
  statutory duty breach of, 74, 75, 76, 77, 78
DEFENDANTS,
  definition of,
    claims for damages in, 7
DISABLED PERSONS,
  occupiers' liability to, 85
DISABLEMENT BENEFIT,
  amount of, 105
  claim for,
    entitlement to, 103
    industrial accident for, 102
    prescribed disease for, 103
DISCLOSURE,
  damages for,
    cases of negligence in, 43
DISEASES,
  claims at common law for,
    nature of liability in, 5
  claims for damages for, 5
  claim in tort for,
    constituents of, 6. *See also* TORT OF NEGLIGENCE.
DUTY OF CARE,
  tort of negligence in, 9, 10, 11

EARNINGS,
  loss of,
    negligence cases in, 45
  loss of future,
    cases of negligence in, 45

EMPLOYEE,
  abnormal susceptibility of,
    standard of care for, 13
  carelessness of,
    employers' liability for, 14
  claims by,
    courts in, 5 *et seq.*
    industrial tribunals in, 96
    salary and wages while incapacitated för, 56
    social security benefits for, 101
  claims for remuneration by,
    suspension from work on medical grounds for,
      cases of negligence in, 58
  definition, 1, 6
  liability for acts of,
    cases of negligence in, 37
  negligence of,
    cases of negligence in, 33
  practical joking by,
    cases of negligence by employer in, 33
EMPLOYER,
  duty of,
    standard of care for, 12
    safe system of work to provide, 11
  liability for acts of others,
    negligence cases in, 37
  standard of care of,
    claims of tort in, 12
  vicarious liability of,
    cases of negligence in, 37
EMPLOYER'S DUTY,
  design of apparatus regards, 29
  negligence cases in,
    negligence of fellow employee regards, 33
  plant and appliances regards, 29
  protective equipment provision of,
    negligences cases in, 32
  safe place of work, to provide, 27
  safe system of work to provide, 24
EMPLOYER'S LIABILITY,
  breach of statutory duty for, 61
  employee's carelessness for,
    claims of tort in, 14
  general practice of industry and, 17
  independent contractor for,
    cases of negligence in, 42
  occupier as, 79
  physical violence of third parties for,
    tort of negligence in, 11
  safety of premises for,
    tort of negligence in, 10

Index

EMPLOYER'S LIABILITY—*cont.*
  statute under, 61
ENCYCLOPEDIA OF HEALTH AND SAFETY AT WORK, 4
EQUIPMENT,
  defective,
    negligence cases in, 30
  manufacturers and suppliers of,
    independent contractors as, 11
  protective,
    negligence cases in, 30
EEC HARMONISATION PROGRAMME, 110
EEC MATERIALS, 3
EXCLUSION CLAUSES,
  damages for,
    negligence cases in, 51

FACTORIES ACT, 2
FACTORY,
  definition, 2
FATAL ACCIDENTS,
  damages for, 87
  liability for,
    action for,
      perpetuators of, 89
    cause of action for, 87
    claim for,
      deceased's children on behalf of, 94
    claim on behalf of deceased's widow, 93
    commencement of, 91
    common law at, 86
    damages recoverable, 92
    interrelation of statutory rights, 95
    statutory change, 86
    survival of causes of actions of, 86
    when an action to be brought, 90
  liability to dependants, 88
  types of damages, 88
FINES, 115
FIRE AUTHORITIES,
  authority of, 114
  concerns of, 113
FLOOR,
  icy surface on, 28
  loose plank on, 28
  slippery duckboard at the site of watertap, 28
  slipping and tripping on,
    negligence cases in, 28

HEALTH AND SAFETY AT WORK ETC. ACT 1974,
  general duties imposed by, 3

HEALTH AND SAFETY COMMISSION,
  Codes of Practice of, 3, 111
  duties of, 111
  functions of, 111
  guidance notes of, 3
  statutory duties of, 111
  structure of, 111
HEALTH AND SAFETY EXECUTIVE,
  duties of, 113
  enforcement powers of, 3
  main organisation chart of, 112
  powers of, 113
  role of, 113

IMPRISONMENT, 115
IMPROVEMENT NOTICES,
  appeals against,
    industrial tribunals to, 115
  serving of, 114
INDEPENDENT CONTRACTOR,
  delegation of duties to,
    occupier by, 83
  vicarious liability for,
    cases of negligence in, 42
INDUSTRIAL TRIBUNALS,
  appeals to,
    prohibition and improvement notices against, 115
  claims at,
    constructive dismissal for, 98
    safety representatives rights for, 98
  claims by employees in, 96
  unfair dismissal claims in,
    employees by, 96
INJURY,
  nature of,
    duty of care covered by, 10
INSPECTORS,
  appointment by, 114
  improvement notices serving of, by, 114
  powers of, 114
  prohibition notices serving of, by, 115
INSTRUCTIONS,
  requirement of, 15
INSURANCE,
  compulsory, 55

LIMITATION PERIOD,
  damages for,
    negligence cases in, 52

MEDICAL EXPENSES,
  loss of,
    negligence cases in, 45

Index

MINES AND QUARRIES,
  legislation on, 2

NEGLIGENCE,
  proof of, 20

OCCUPATION,
  actual control as, 80
  control exercised through another as, 80
  premises or structures of,
    definition of, 80
OCCUPIER,
  common duty of care of, 81
  definition of, 80
  delegation of duties,
    independent contractor to, 83
  employer as,
    liability for, 79
  employers' liability as,
    scope of 1957 Act in, 79
  exclusion of duty of, 82
  performance of duty of, 83
  warning by,
    visitor to, 83
OCCUPIERS' LIABILITY,
  application to Crown of, 84
  definition of visitors for, 80
  disabled persons to, 85
  trespassers to, 84
OCCUPIERS' LIABILITY ACT 1957,
  scope of, 79
OFFICES, SHOPS AND RAILWAYS PREMISES,
  regulation on definition of, 2
ORDINARY RISK OF SERVICE, 19

PAIN AND SUFFERING,
  damages for,
    negligence cases in, 46
PENALTIES,
  non-compliance for,
    statutory duties of, 115
PERSONAL INJURIES,
  claims at common law for,
    nature of liability in, 5
  claims for damages for, 5
  claim in tort for,
    constitutents of, 6. *See also* TORT OF NEGLIGENCE.
  damages for,
    interest on, 50
    tort of negligence in, 43, 44, 45

PLACE OF WORK,
  employer's duty to make safe,
    cases of negligence in, 27
PLANT,
  employer's liability regards,
    latent defects, 29
PLANT AND APPLIANCES,
  tort of negligence in, 29
  latent defects of,
    negligence cases in, 29
  patent defects of,
    negligence cases in, 29
PNEUMOCONIOSIS, BYSSINOSIS AND ALLIED DISEASES,
  prescribed diseases as, 103
PNEUMOCONIOSIS, ETC. (WORKERS' COMPENSATION) ACT 1979,
  payment under,
    lung disease for, 105
PNEUMOCONIOSIS, ETC. (WORKERS' COMPENSATION) (PAYMENT OF CLAIMS) REGULATIONS 1985,
  lump sums payable under, 106
PRACTICAL JOKING,
  employee by, 33
PRESCRIBED DISEASES,
  classification of, 103
  disablement benefit for, 103
PROHIBITION NOTICES,
  appeals against,
    industrial tribunals to, 115
  serving of, 115
PROOF,
  tort of negligence for, 20
  negligence of, 20
PROTECTIVE EQUIPMENT,
  defintion of, 32
  provision of,
    negligence cases in, 32
PROVISIONAL DAMAGES,
  cases of serious deterioration in, 44

REMOTENESS,
  damnages, of,
    cases of negligence in, 44
RES IPSA LOQUITUR,
  cases of negligence in, 21
ROBENS COMMITTEE ON SAFETY AND HEALTH AT WORK,
  report of, 3
ROOF WORK,
  place of safety at work cases, 28
RULES OF THE SUPREME COURT,
  damages on, 44

SAFE SYSTEM OF WORK,
  employers' duty to provide, 11
SAFETY, HEALTH AND WELFARE
    LEGISLATION,
  administration and enforcement of,
    110
  compliance with,
    criminal law by, 115
  enforcement powers of, 114
SAFETY OF PREMISES,
  employers' liability for,
    tort of negligence in, 10
SAFETY REPRESENTATIVES,
  rights for time off, 99
  rights of,
    claim for,
      industrial tribunal at, 98
    right to be consulted, 98
SALARY AND WAGES,
  employees' claim for,
    incapacitated while,
      cases of negligences, in, 56
SELF-EMPLOYED, 5
SERVANT,
  definition, 1
  loan of, 41
SICKNESS AND INVALIDITY BENEFIT,
  contribution record for, 105
  entitlement to, 105
SOCIAL SECURITY,
  damages and,
    negligence cases in, 48
SOCIAL SECURITY APPEAL TRIBUNALS,
    107
  legal aid for representation before,
    108
SOCIAL SECURITY BENEFITS,
  claims by employees for, 101
  system of adjudication for, 107
SOCIAL SECURITY COMMISSIONER, 107
SOURCE MATERIAL,
  Acts and Regulations, 2
STANDARD OF CARE,
  danger unknown at time of accident,
    17
  general practice of industry, 17
  magnitude of risk, 12
  ordinary risks of service, 19
  relevance of previous accidents, 16
  statutory duty, for, 68
  supervision and enforcement
    precautions, 16
  young people for, 15

STATUTORY DUTY,
  breach of,
    act of plaintiff caused by, 75
    contributory negligence as a
      defence for, 77
    damages for, 78
    defences for, 74
    definition of who can sue for, 62
    employer's liability for, 61
    latent defect as defence to, 75
    rules of construction for, 6
    volenti non fit injuria, 77
    who is liable for, 64
  class of person authorised as owed, 64
  contracting out of, 74
  defence to breach of,
    act of independent contractor as,
      75
  delegation of,
    plaintiff to, 74
    third party to, 75
  independent contractor, owed to, 64
  limitation of,
    particular class of injury for, 67
  owed to,
    employer by, 62
  "person employed in the process"
    owed to, 63
  "person employed" owed to, 63
  proof of breach of,
    cases where duty to provide
      safeguard, 73
    onus of, 72
  standard of care for,
    meaning of danger and dangerous
      for, 68
    meaning of efficient for, 71
    meaning of maintenance and repair
      for, 72
    meaning of practicable and
      reasonably practicable for, 70
    meaning of provided and available,
      71
    meaning of safe for, 68
    meaning of secure and securely,
      69
STATUTORY LIABILITY,
  jurisdiction for, 65
  occupiers, of, 64
STATUTORY PROTECTION,
  principal areas covered by, 61
STATUTORY PROVISIONS,
  enforcement powers of, 114
STATUTORY SAFETY CODE,
  workings of, 62

Index

STATUTORY SICK PAY,
  calculation of, 57
  claims for,
    cases of negligence in, 57
SYSTEM OF WORK, 24
  duty of employer to provide,
    cases of obvious danger, 25
  employers duty to provide,
    time of, 24
TAXATION,
  damages and,
    negligence cases in, 48
TORT OF NEGLIGENCE,
  acts of agents in, 37
  claims by employees,
    statutory sick pay for, 57
  compulsory insurance in cases of, 55
  constituents of, 6
  damages,
    disclosure of documents in, 43
    dismissal for want of prosecution, 54
    events after the accident for, 47
    exclusion clauses for, 51
    interest on, 50
    limitation period for, 52
    loss of amenities in, 46
    object of, 43
    pain and suffering for, 46
    quantification of, 43, 45
    remoteness of, 44
    social security and, 48
    taxation and, 48
  defective equipment in, 30
  defences in,
    contributory negligence, 34
    volenti non fit injuria, 36
  defendant in, 7
    claims for contribution or
      indemnity against, 8
    Crown as, 7
    Foreign States as, 8
  duty of care in,
    general scope of, 9
    nature of injury, damage etc.
      covered by, 10
    place where owed, 10
    special susceptibility of employee
      when, 9
  duty of employer in,
    negligence of third parties in
      respect of, 10
  employee's acts in, 37

TORT OF NEGLIGENCE—*cont.*
  employee's claim for remuneration
    in,
    suspension from work on medical
      grounds for, 58
  employee's claims in,
    salary and wages while
      incapacitated for, 56
  employer's liability in,
    physical violence of third parties
      for, 11
    safety of premises for, 10
  interim payment of damages in, 44
  liability for acts of others in, 37
    course of employement in, 38
    independent contractors, 42
    loan of servant, 41
  loss following,
    financial of non-financial, 45
  loss of future earnings in, 45
  negligence of fellow employee cases
    in, 33
  place of work, 27
    means of access at, 27
  plaintiff, 6
  plant and appliances in, 29
    latent defects of, 29
    patent defects of, 29
  practical joking by an employee in, 33
    causation in, 22
  proof of negligence in,
    generally, 26
    how the accident happened for, 20
    res ipsa liquitur, 21
  protective equipment,
    provision of, 30
    what constitutes provision of, 32
  standard of care,
    abonormal susceptibility of
      employee for, 13
    assessing the magnitude of the risk
      for, 12
    carelessness of employes and, 14
    danger unknown at time of
      accident, 17
    devising reasonable precautions
      for, 12
    foreseeing the existence of a risk
      for, 12
    general practice of industry for, 17
    main heads for, 12
    ordinary risks of service and, 19
    relevance of previous accidents and
      complaints for, 16
    requirement of instructions and, 15

TORT OF NEGLIGENCE—*cont.*
  standard of care—*cont.*
    statutory safety provisions
      relevance to, 18
    supervision and enforcement of
      precautions in, 16
    work on premises of third party,
      for, 19
  system of work,
    cases involving lifting heavy
      objects, 26
    cases of obvious danger, 25
    employer's duty to provide, 24
    windows cleaning cases, 25, 26
  vicarious liability,
    criminal acts for, 40
    loan of a servant cases, 41
TRESPASSERS,
  occupiers' liability to, 84

UNFAIR DISMISSAL,
  claim for,
    industrial tribunal in, 96
    definition of, 97

VICARIOUS LIABILITY,
  course of employment during, 38

VICARIOUS LIABILITY—*cont.*
  criminal act for,
    during course of employment, 40
  employer of,
    cases of negligence in, 37
  independent contractors for,
    cases of negligence in, 42
  loan of a servant cases for, 41
VIOLENT TREATMENT,
  duty of care to prevent, 11
VISITORS, 1
  definition of,
    liability owed to, 80
VOLENTI NON FIT INJURIA,
  defence as,
    cases of negligence in, 36
    statutory duty breach of, 77

WINDOW CLEANING,
  dangerous operation of, 25
WORK,
  places of,
    tort of negligence in, 28
WORK PLACE,, 27

YOUNG PEOPLE,
  standard of care owed to, 15